Enthusiastic endorsements for
FINDING CARLA

Ross Nixon is a fantastic storyteller and knows a fantastic story to tell. *Finding Carla* is a captivating tale about aviation judgment, the will to survive and the aviation community's enduring ambition to make flying safer for all. If you'd like to know why airplanes today carry emergency locator transmitters, or just have a need to be kept on the edge of your seat with an engaging story, then you must read *Finding Carla*.

Rod Machado—Author, Speaker, Flight Instructor

I just finished the draft of your book, *Finding Carla*. I am so impressed with your ability to put the facts, the testimony of others, your own thoughts and aviation expertise into an important story. I could not put it down!

Johnny Moore—Author, FAA Wright Brothers Master Pilot 2015

An enthralling, exceptionally well written and researched book—an epic story of air crash survival and death in the wilds of Northern California... I recommend this book without reservation. A must read for every pilot and layman alike.

G. Pat Macha, www.aircraftwrecks.com

"A real page turner. Thank you Oien family for saving thousands of lives. Your pain and suffering was not in vain."

—Jim and Ferno Tweto, "Flying Wild Alaska"

FINDING carla

The story that forever changed
aviation search and rescue

Ross Nixon

Ross M. Nixon

Aviation Supplies & Academics, Inc.
Newcastle, Washington

Finding Carla
The story that forever changed aviation search and rescue
by Ross Nixon

Aviation Supplies & Academics, Inc.
7005 132nd Place SE | Newcastle, Washington 98059-3153
asa@asa2fly.com | www.asa2fly.com

ASA-FIND
ISBN 978-1-61954-343-0

Printed in the United States of America
2020   2019   2018   2017   2016   9 8 7 6 5 4 3 2 1

Library of Congress Cataloging-in-Publication Data:
Names: Nixon, Ross.
Title: Finding Carla : the story that forever changed aviation search and rescue / by Ross Nixon.
Description: Newcastle, WA : Aviation Supplies & Academics, Inc., 2016.
Identifiers: LCCN 2016005074| ISBN 9781619543430 (trade paper) | ISBN 1619543435 (trade paper)
Subjects: LCSH: Aircraft accidents—California, Northern—History—20th century. | Corbus, Carla, 1951-1967—Death and burial. | Corbus, Carla, 1951-1967—Diaries. | Oien, Alvin F., 1908-1967. | Oien, Alvin F., 1908-1967—Family. | Airplanes—United States—Radio equipment—History—20th century. | Aircraft accidents—United States—Law and legislation. | Airplanes—United States—Radio equipment—Law and legislation. | United States. Federal Aviation Administration—History—20th century. | United States. National Transportation Safety Board—History—20th century.
Classification: LCC TL553.525.C2 N59 2016 | DDC 363.12/47—dc23
LC record available at http://lccn.loc.gov/2016005074

Contents

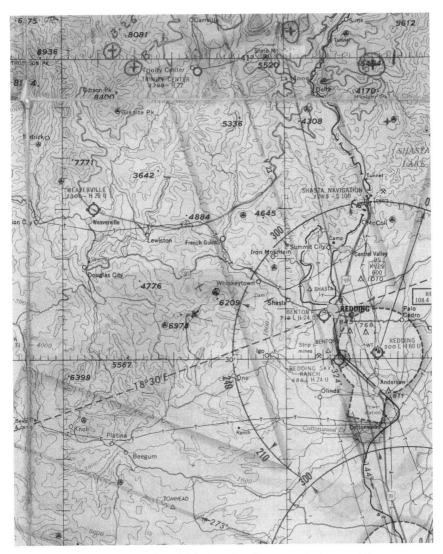

From the well-pored-over, often-folded sectional-aeronautical map of the
Trinity Mountains area of Northern California that the Oien brothers used while
searching for their dad's downed Cessna 195. Al Jr. indicated in pencil several
other wrecks he found by circled-airplane symbols.

Preface

The ELT Beacon Law

Ever since the Code of Hammurabi was scribed onto stone, people have said, "there oughta be a law about that!" We now live under the rule of law and there are more laws now than Hammurabi could possibly chisel into stone. Sometimes we feel a new regulation is just another mandate being jammed down our throats, but when the lawmakers name the law from the inspiration for it, that makes it easier to see the humanity behind it: the Lindberg Law, the Adam Walsh Child Protection Law, and Kristen's Law are all such examples of rules that came to be from tragedies lived out by individuals.

The ELT beacon law, signed into the regulations in 1970 as a rider to the OSHA bill, could well be called Carla's Law. In the 1960s there had been a push for ELT beacons to be installed in aircraft in the USA, but in the halls of our nation's Capitol the political will to mandate them aboard aircraft did not exist. Though it would save millions of dollars in SAR costs and save lives, the usual suspects that detract from progress interfered: expense, interruption of the status quo, and political gamesmanship.

It appalled Senator Pete Dominick of Colorado, a flyer himself, that a simple radio beacon that could pinpoint crashes was not mandatory equipment aboard U.S. aircraft. There was a list of accidents read by an increasingly frustrated Senator into the congressional record. When the details of the "Carla Corbus accident"—the events of this story—made headline news, it was the shocking straw that broke down all political resistance. It was almost as if fate used Carla and her family to make a terrible point to Pete Dominick's colleagues. In a few short years, ELT beacons, though at that point imperfect, were mandated to be carried onboard all U.S. civil aircraft.

This is a flying story as well as a life story. I pass no judgment on the pilots mentioned in here because I too have worn out a luck charm or two while flying. In retelling this story, my hope is that the messages of this book spread through the aviation world and beyond, and cause people to think about what is important in flying and life—and maybe even about how they want to be remembered. The Oien family's sacrifice in this needs to be remembered.

Acknowledgments

There are so many people who helped me along the lonely writer's path. The story was put to page in pilot quarters all over Alaska, in stark places like Nome, Kotzebue, St. Mary's, Bethel, and Barrow. The guys and gals I flew with read portions, knowing I could fly, but probably wondered if I could write. Here is my proof and I thank all my friends who read my works and gave me input.

Two editors from my adopted hometown of Anchorage encouraged me: Rebecca Goodrich and David Holthouse. The ASA editors, Jennie Trerise and Jackie Spanitz, were awesome, too, taking my scribble and turning it into a book. Best of all, being a writer excused me from a lot of man chores at home. My wife Kate often said: "Just write"...and so I did.

None of us would be anywhere without our moms and I'd like to thank mine, Viola Nixon. When I visited Al Oien at his home in Washington, which happened to be my hometown, he and his wife Carol graciously invited me to stay with them, but I always wanted to stay at home, the place mom kept together after an air tragedy struck in our family. Though jolted hard by the loss of our dad, five great kids spring-boarded from that house full of books and ideas adding value to everything we've done.

I thank the Oien boys too, sons of the strong-willed man you will read about, who stepped up to the plate when the big question mark of a missing plane and family came over their lives. They shared their story of how they put aside their feelings towards a man who'd been so hard on them. They took care of business, living up to their obligations with honor. They did not give up hope because: "He was our dad," they said. The final outcome was beyond their control, but because of the ripple effect of their troubles and loss, aviation is a safer endeavor.

*This book is dedicated to
my wife Kate, to my mom, Viola;
to the Oien boys: Ron, Chuck,
and Al, Jr., and to Carol Oien.*

FINDING carla

A man and his plane: Al Oien, Sr. and his Cessna 195, "The Viking"

The author as a young boy (at left) works on the family Cessna 195 with his brother and sister, Matt and Ingrid, circa 1967.

Hangar Flying
Or, The Beginnings of a Story Retold

T HE DEER HUNTERS FOUND this note on top of the stack of letters:
"*Whoever finds this wreck*

Please mail these letters for us. We waited so long for you. Where were you?

Our daily log is here for you to see in the folded airman's guide.

PLEASE MAIL THESE LETTERS"

I'd sought these mysterious messages for almost forty years. My hands shook and heart thumped as I held them. A lifelong, personal mystery was solved.

It all began when I was a boy. In my family we never went to church. We went to the airport instead. Both my parents flew planes. Although he was one, Dad did not resemble a member of the American College of Surgeons when at his hangar. He was a big strapping guy who wore overalls and puffed on Swisher cigars when he worked on his planes, while I rolled around on my back on a mechanic's creeper scrubbing the airplane bellies clean with Formula 409 and old rags. As the cleaner dripped into my eyes, I dreamt of the day I'd fly a plane.

Dad had a bunch of planes. They were the type of planes a welfare kid from the 1930s purchased once he made some money. The crown jewels were the Staggerwing Beech D-17S, and a pristine Cessna 195. That elegant Staggerwing took up most of my polishing time with her oil and smoke-belching Pratt and Whitney 450 HP engine. In spite of her dirty ways, she was a great classic. In her day, in the '30s, she out-flew frontline U.S. military fighters. Mom flew a Cessna 170 and I helped restore a Cessna 140 that I soloed at age 16.

A weak, sickly kid, I loved the Beechcraft but I preferred our Cessna 195, a fast, all-metal bird with handsome lines. The 195

"Business Liner" embodied the way I sought to be but was not: strong and solid. I worked cleaning her blue and white Alumagrip paint because someday the plane would become mine, or so dad said.

My father had been a logger, a tugboat man, and a fireman on a steam locomotive; the last one on the Canadian West Coast, he'd remind you. When he began his medical practice in Canada he was a bush flying doctor. He landed his Stinson Voyager near your place on wheels or skis, then walked up to your front door with his medical bag in hand. He was one of those old school pilots who learned to jump in a plane, point the nose and go.

He got good at flying doing that sort of work so when he spoke I soaked up the words. In aviation there is an art of storytelling known as "hangar flying," occurring wherever pilots gather and talk. Dad's stable full of planes provided the local pilots a perfect place to hangar fly. His friends saw his dingy 1962 Pontiac airport car parked outside and stopped there to "shoot the bull" with Doc Nixon, while I rolled around the hangar floor, ignored like Chief Broom from *One Flew Over The Cuckoo's Nest*.

I see Dad now on that particular day during the early '70s at the hangar, standing in his overalls puffing on the cigar, talking solemnly with some of his pilot friends. In his hands he held an orange plastic box. It looked like a radio, but carried no antenna or speaker. Through puffs of Swisher cigar smoke he explained how the box housed an emergency locator beacon, or transmitter (an "ELT"). Impact forces triggered a switch activating the radio, which sent out a distress signal. He showed them the flexible whip antenna on the fuselage of the Cessna 195.

He followed this demonstration with the tale of a family who died in the wilderness, down in the "Siskyous." The people survived the wilderness crash of their own Cessna 195, but slowly wasted away after living for months in the wreckage. They left behind a diary and series of letters. Because of this event, the FAA passed laws requiring these emergency locator transmitters to be put aboard all U.S. civil aircraft. Dad installed the ELTs in all four of his planes.

The tale haunted me and I could not shake my visions. Bleak mental sketches in artist's gray pencil imagined a Cessna 195 wrecked on a mountainside, a man walking through thick, snowy brush and a mother and her daughter at the crash, slowly starving. I saw images of handwriting. The doomed family lived in my mind, always in the bleak colors of gray, white, and black. Questions ran through my head. When did this happen? Where and to who? Terrible indeed, but the sort of thing that happened to other people, not us.

Dad was a hell of a pilot. He held a world flight distance record and had won the Bleriot Medal, but he proved me wrong by going down when I turned seventeen years old, leaving my mother and four siblings with broken hearts and bills. He'd not been the most sensible money manager and our affluent ways went down with his plane. One by one the planes were sold so we could live. The classic Beechcraft went off to Canada. My beautiful 195 flew off with some dentist. Time healed the wounds of our derailment and eventually I got back on track.

For years after I'd stood in that western Washington hangar, I wondered about the lost family of the Siskiyous. Who were they? When did they crash? Did it really even happen? As a man I flew commercial planes. At work I taught pilots about the orange rescue radios known as ELTs. One even saved me from some lonely hours after a mishap on the Alaska tundra when I misjudged distance during an off airport venture, and my Piper bush plane flipped upside down in thick tundra grass. I switched on the portable ELT and prepared for a lonely night. It wasn't long until my flying buddies found me, landed nearby, and helped right my damaged bird onto her wheels. Later, safe after my stupidity, I silently thanked the long-lost family who inspired the push to include the radio technology. I now owed them. Their ghosts lived in my head. Again the questions rolled through my brain... Who were these people? Did they ever really exist?

Over the years in libraries I'd checked the reader's guidebooks for periodical literature for articles about a marooned family in the Siskiyou Mountains without any luck. Later on, with the Internet available, I searched again for details on a Siskiyou-area plane crash involving marooned people who wrote a diary, and found nothing.

The mystery family tugged at my mind like the fable of some lost gold mine. One day I came across a note in a flying article referencing the "Carla Corbus death diary." When I saw those words, I knew I'd finally found the key to the old hangar tale. Yes, it actually happened. Feeling like I struck an Internet mother lode, I stayed up late Googling Carla Corbus and her diary, and reading the details of the tragedy.

I felt I'd seen it all before, viewing the black and white newspaper photos of the crash victims peering back at me from the computer screen. The pilot, Alvin F. Oien, Sr., carried the same confident grin of my father. Carla Corbus and her mother Phyllis looked straight to my heart from the ancient pages. As I read through the news stories, the horror of it all struck me. Though their demise was torturously slow, I noticed a brighter side to the story, too—the epilogue known by few.

Links to stories of people who'd been saved by ELT radios appeared. I saw the suffering of Carla Corbus indeed directly tied to the advent of the ELT rescue radio. Because of the suffering of three forgotten people on a forgettable California mountainside, thousands of others lived.

I noticed a brighter side to the story, too—the epilogue known by few.

Dad was incorrect about the crash location being the Siskiyous, a mountain range on the Oregon-California border. The initial news reports even erred about the site, calling it by the weird name "Bully Choop Mountain." The family suffered their lonely end in a range called the Trinity Mountains, just outside of Redding, California, on a peak with the odd name of Shoemaker Bally, right next to Bully Choop Peak. The accident happened in March 1967 and the diary that turned up on top of the wreck's instrument panel shocked the world.

Knowing the truth, the story haunted me worse than ever. I dug deep, discovering a 1968 *Saturday Evening Post* article on the crash. Admirably, the pilot's son spent months flying over the rugged mountains of Northern California long after the official search and rescue efforts ended. He and his brothers never gave up hope. I wondered if this man still lived. He'd be an old pilot now because he was in his early 30s at the time of the accident.

I found the man through the Internet, coincidentally living only a few miles from Dad's hangar where I first heard of the rescue radio during that hangar flying session through the clouds of cigar smoke.

I emailed Alvin F. Oien, Jr., asking if he knew about a crash in Northern California back in 1967. He replied to the message in all capitals: I AM THE PILOT'S SON, WHAT CAN I DO FOR YOU? From the articles I knew he'd been a military pilot-officer, then an airline pilot. I pegged him for an old-school man and knew there could be no nonsense when dealing with him, although his emails were signed, "Smiling Al."

After writing to him of my interest in the crash story, Oien Jr. replied that he would provide information. I emailed him questions. I read more articles, made calls, and searched archives. In time my inquiry *I AM THE PILOT'S SON, WHAT CAN I DO FOR YOU?* took me to the crash site, to the woman who found the wreck, to the hotel once owned by the pilot Alvin F. Oien, Sr. in downtown Portland.

I became good friends with Oien Jr. and visited him at his house in my old hometown. He was in his mid 70s when I met him. He arrived there as a retiree about the same time I graduated high school in the bleakness of western Washington in 1979. With a newly deceased father, a tanked economy, and few prospects, life did not look rosy for me at the time, though for him the low property values were a dream.

Oien's Olympic Peninsula home was a man's place, with big wooden beams, huge windows, and solid brown furniture. It was loaded with the memorabilia of a world-traveling pilot's life. His 1950s military flying career amazed me: the last B-17s in the USAF inventory to C-54s and the Convair C-121. At Delta Airlines he flew the C-46 freighters then retired from the Boeing 727. He'd even flown the Atlantic in his own Piper Apache. He was burly and gruff. He wore overalls claiming they were all he ever wore and carried a Texas accent from living down there most of his life, though he was from the Northwest.

His property sat on the edge of the Elwha River valley, where he lived for part of the year. Outside, on that summer morning when we met, the Northwest fog obscured the sun, but through breaks above you'd glimpse the towering Olympic Mountains shining in sunlight. Placed right on the glide path to the William Fairchild International Airport, where my own dad departed on his last flight and where I learned to fly, the inbound planes flew directly over the house. As we sat in the kitchen, we'd guess as to the type of plane buzzing above. Al's wife Carol hovered over us, filling our coffee cups.

Al had not forgotten the crash and search for his father's family. To prepare for my visit, he'd laid out the artifacts he still lugged around from 1967, saying the suitcases were packed in his garage, just the way they were found on the mountain. His kitchen tabletop held stacks of magazines, piles of photos, papers, and maps, as well as various small curious objects. Among the items there I saw the death diary, recognizing its first page from news photos. Part of me wanted to grab it, to rip through the words I'd wondered about for so long…but I knew I'd opened some long-closed wounds by digging into this event. I'd be patient.

Al said, "I've not looked at these things for forty years!" as he picked up various items, looked at them, and put them down very slowly and carefully. We sipped black coffee and listened to the planes pass above. I just let him talk when he felt like it.

"Here is the case for the milk of magnesia tablets they ate when they ran out of candy. Here is one of Phyllis's gloves." He picked through the tabletop items. He held a tan woman's dress glove.

"Here is the will that the Old Man," as Al referred to his dad, "wrote while dying on the mountainside."

We looked at the deteriorated paper, which was barely holding together. The paper was a receipt from Rupert Flying Service, Beaverton Airport, Oregon.

Dying, Al Sr. had written his will on the back.

"You can see how the animals chewed the wallet edges." He held out a cheap plastic wallet emblazoned with the name of the Clifford Hotel, in Portland, Oregon. Once rectangular, now oval, chewed into that shape from the teeth of small rodents.

"He used to give these away to guests. I've got a million of them in the garage."

I picked out the faded words from the scrap of receipt: "PLEASE SEND HELP!" with barely legible requests for burial next to his daughter (the one lost from a childhood illness), and requests about his estate for his attorney and accountant. There was more, but it was indecipherable.

These were not the words of a man who abandoned his family, as some articles had seemed to suggest. I could see Al's father struggling to write these last words with his left hand, his right arm useless and broken, once he realized he'd never make it out of the wilds alive… All this writing, after struggling through armpit-deep snow down a steep mountainside in a futile effort to get help.

Al picked through some curious rectangular silver-colored pieces of Naugahyde. "These are the playing cards they made from the upholstery. It looks like the ink has faded. Each one used to carry a number and suit. Remember how the article said how they played pinochle to pass the time? These are their cards right here."

I held some of the tiny cards, chilled by the fact they'd been made and played by Phyllis Hausheer Oien and Carla Corbus on that cold, cold mountain while the two waited for rescue. Al gave me one of the cards, which I carried thereafter for writing inspiration. There were stacks of photos, sympathy cards, and magazines from around the world. I spied a photo of a young girl and recognized her immediately as Carla. I'd studied her face as it stared at me from the news pages I'd found on Google. Now she peered at me from a sofa, the picture dated February 1967, one month before the accident. If only I could have warned her not to go on the flight…she'd be sixty-plus years old had she survived.

We skimmed through the diary together, reading the day-by-day account of a long, slow decline written on the margins of the Airman's Guidebook. Al knew each word because he'd studied it intensely in 1967 and 1968, trying to find clues to the location of his father, who'd tried walking out of the mountains. The words laid out the tale of a short, bitter life on the mountain from the beginning to the end. Like the diary of Anne Frank, these were messages from a grave. I'd wondered about this document for many long years and now I held it in my two hands, the mystery solved. In aviation training, instructors always claimed the rules were written in blood. This tabletop was covered with the blood behind the lifesaving ELT beacon regulations.

I puzzled over a small aluminum door that lay among the tabletop goods. Something inside me recognized the distinct plastic knob. Al said, "It is the only piece of the plane I own now." Then it struck me. I'd stared at and fiddled with a door just like that when I flew in the right seat of Dad's Cessna 195, back when I was the weak kid with dreams of owning that particular plane. It was a glove box door from the crashed Cessna. I once knew that little door like the back of my hand, positioned right smack in front of the copilot seat of my youth.

I went on my way, back up north to Alaska. As I sat on the jet, I wondered what led me to these days with Al Oien? I could easily have been elsewhere long ago when Dad told the tale during that hangar flying session. Why were my bleak visions so accurate? Had

I seen the papers as a six-year-old boy and heard someone discuss the crash? Or are we really connected in ways we don't yet understand?

Parallels existed between our fathers: Both were successful men who'd risen above poverty, both ex-loggers, both flew Cessna 195s, both men were capable pilots who flew off into oblivion. Both smiled like they owned the world and in a way they did while they lived, but they lost it all through exercising the bold decisive-ness that guided their lives. Then, of all things: my being saved by an ELT years later. Lastly there existed the fact of the pilot's son retiring and keeping the enig-matic documents so close to the hangar at the airport from where I learned to fly and where I first heard the story from my father during that long past hangar flying session.

Parallels existed between our fathers: Both were successful men who'd risen above poverty, both ex-loggers, both flew Cessna 195s, both men were capable pilots who flew off into oblivion.

When I resolved to write this tale for you, the ancient voice of the boy crept into my head. *"You! You're not a real writer,"* but in the end the proper muse appeared. In its own way this story is the "Perfect Storm" of the aviation world. On his creation of that great sea yarn, Sebastian Junger wrote how he started on his work, an outsider looking into the facts. He perse-vered: meeting people, collecting the data, making notes, assembling it. In time the story became part of him. The same occurred here, with me. Although none of what follows happened to me, this writ-ing has become a big part of my heart. That is how the old hangar flying tale from my boyhood became the story you are about to read.

CHAPTER 1

A Warning Shot

Oregon, Friday March 10, 1967

THE GREEN AND WHITE West Coast Airlines Fairchild F-27, bearing U.S. registration number N2712, touched down at Klamath Falls Airport just before midnight on March 9, 1967. Like most airlines, West Coast Airlines shuffled its fleet according to profitability, not employee comfort. After minimal rest at a local motel, the two pilots and the flight attendant returned to the airport for their Seattle run.

A few hours of sleep is never enough for a day's flying. Composing the crew were Dale Anderson, a 4,000-hour captain who served as pilot-in-command; Tom Zeiders flew in the right seat as co-pilot; Connie Berryman, 22, a new employee with just over six months flying for West Coast, took care of the passengers, serving as the flight attendant.

For the previous few days, the U.S. Aviation Weather Service warned aviators of icing conditions aloft in Southern Oregon and Northern California. Now, five hours after their late arrival, these people were slated to fly the airliner back up through the same icy clouds they'd penetrated the evening before while landing at Klamath Falls.

Offshore to the west lay a colossal low-pressure storm system. A thousand mile long cold front shot from the center of the low and spanned from Canada to Mexico, with a small barb pushing inland and nearly stalled near the Oregon-California border, right over the Klamath Falls airport. Weather charts for the forecast depicted the cold front as a line studded with sharp points spinning eastward from the center of the low. Where the line sliced the map, the border country suffered under a blitz of rain, wind, and snow. For now, dismal conditions prevailed at Klamath Falls, coating everything with a soggy layer of snow mixed with rain. Fortunately the Fairchild airliner spent the night in the West Coast Airlines hangar, otherwise there'd be the big job of deicing before the departure.

The tired crew rolled back to the Klamath Falls airfield at 0400 in a taxi. Most pilots look at the sky on the way to the airport. If they were like most, Anderson and Zeiders noted the dismal air during the morning ride.

To get out of Klamath Falls, they'd punch upwards through the icy clouds. Once they flew high enough they'd be up in the moonlight and clear air. They did not anticipate any trouble.

Their Fairchild F-27 carried ice removal equipment. Heated propellers and rubber bladders called "boots" on the wings and tail helped keep those critical surfaces clear of performance-degrading ice in flight.

No doubt at 0400 the crew thought ahead to the end of their long day and then getting on with their weekend. The three prepared the Fairchild for the day's flying.

During the night, mechanics inspected the F-27. When they completed the work, they signed the aircraft's log assuring no mechanical discrepancies existed. Anderson re-checked the flight's plan with a dispatcher. Both agreed on a 0500 departure. The dispatcher released the flight and the Captain signed the necessary form.

Usually the turboprop carried only a few passengers north from Southern Oregon. As it stopped at the cities en route to Seattle, the 40 seats filled. When Anderson and Zeiders took their places in the cockpit, just one man took a seat in the back.

Boarding the sole passenger inside the hangar, though unusual, saved precious time. If the plane taxied outside to the terminal, parked, then loaded, snow or sleet could foul the wings. Then there'd be a delay while the ground crews sprayed the deice fluid. The two pilots reasoned their plane could be tugged out from the hangar. They'd quickly fire up the engines and get airborne before any lift-destroying ice formed on the ninety-foot long wings.

The captain of any aircraft is known as the pilot-in-command. He has the authority and responsibility to conduct the flight safely, and if anything goes wrong, he is at fault. Captain Anderson carried the final word on deicing. If he commanded it, ground crews would spray the wings with expensive deice fluid until they were clear of contamination.

While the two airline pilots waited for the tug to pull their airliner from the heated hangar, a private aircraft up near Portland also sat prepped for flight in a hangar at the old Beaverton Airport. This other plane, a small Cessna 195, sported gray paint with blue trim and a red rose on the tail. The right side entry door carried the name of the plane: THE VIKING. The Viking's owner planned to fly the little Cessna to San Francisco the next morning where he, along with his wife and daughter, would board a Dallas-bound Delta Airlines flight. The Cessna's planned route crossed the icy gauntlet of the southern Oregon border region.

Over the week, this pilot carefully serviced the little plane, loading it with extra oil and tools. Later he'd pick up three full suitcases from his house. When he topped off the fuel tanks, he'd put the suitcases aboard. Hopefully there'd be enough room for the survival kit. But, if necessary, he'd leave it behind. The route to California lay mostly over civilized country with roads. Airports were plentiful between Portland and the Bay Area, except for one small stretch across Northern California.

The big Fairchild airliner in Klamath Falls cranked the Rolls-Royce turbines and taxied for departure. The pilots conversed. The only known private communications from the pilots came from the cockpit voice recorder (CVR), which survived the accident.

The cockpit voice recorder is known as the "black box," though it is actually orange. In airliners, the CVR is designed to withstand fire and severe impact. It automatically records the last ninety minutes of cockpit conversation and starts recording automatically at the beginning of a flight.

Problems began for Flight 720 right away. Anderson and Zeiders both discussed how the snow stuck to the airframe as the plane cleared the hangar. The warm plane moving into the colder outside air caused the falling snow to melt and then refreeze. The pilots commented twice on the plane's icy wings as they taxied.

A flyer finds it much easier to make a firm and final decision about flight during horrible weather conditions; it is said the marginal conditions are the ones that kill you. Zeiders and Anderson faced those marginal conditions. Their dilemma really was not a question of flying or not, it was the question of whether to take time to deice and delay the flight.

The sleety conditions skewed Anderson's judgment or maybe other pressures such as keeping the flight on time kept him from thinking clearly. Maybe fatigue played a role. Maybe it was the thousand or so hours he had flying the DC-3, a plane known to fly well with ice. Against the Fairchild's aluminum skin the falling precipitation seemed to melt into harmless rain.

The pilots commented twice on the plane's icy wings as they taxied.

The morning proved to be a fiasco. Wet snow on the ramp mired the tug, taking ten minutes to get rolling again. The Fairchild's wing collected the mush then it melted away, while the ground crew dug out the tug.

Anderson kept looking at the wing, perhaps only seeing what he wanted to see. He'd worry more if the plane carried a full load. The lighter the load, the shorter the takeoff. Once airborne, the deicing equipment could help eliminate any ice, or it could just sublimate off the wings and disappear back into the black air. Over the intercom Berryman reported all ready in the cabin.

As Flight 720 taxied toward the runway, Zeiders copied the departure clearance, releasing the Fairchild to 10,000 feet onto the airway for Medford, and advising it to contact departure control on frequency 124.1 once airborne.

The cockpit voice recorder captured Zeiders saying, "It is starting to freeze outside." Anderson replied. He "did not care if we were below freezing. They did not put on any isopropyl on here and the stuff sticks," referring to the fluid used to prevent ice build-ups on aircraft.

The control tower gave a final weather brief, reporting the visibility as one mile, sky obscured with a ceiling of 700 feet, overcast with light snow. The two pilots completed the pre-takeoff checklist with Zeiders reading it and Anderson responding. At 0501, lining up on Runway 14, Anderson put the throttles forward on the console, putting full power to the Rolls Royce engines, saying, "We gotta go before that stuff freezes!" He succeeded. The flight left on time. Zeiders responded "Yeah," on the cockpit voice recorder.

The Fairchild leapt forward onto the dark runway, lit only by the edge lights to the left and right. Inside the comfortable cockpit, Zeiders called out the instrument readings, his face reflected by the comforting glow of the instrument lights.

A cockpit provides great refuge in glum weather. Warm, dry, and with all the knobs and controls so familiar, they are like snug second homes to pilots. Sometimes a plane seems so big, reliable, and powerful that nothing can go wrong. Like those running the Titanic, flyers sometimes fail to recognize their vulnerability. To get the F-27 moving, to get going on their day, relieved both pilots. Not deicing helped them keep on schedule. Maybe Anderson felt he'd made a good call, saving time and money for the company.

Perhaps the Fairchild's controls felt strange to Captain Anderson, but all the instruments read normally as the plane accelerated. Zeiders, being the pilot monitoring called nothing out of the ordinary.

There is a moment on each takeoff when a pilot may safely abort. If anything abnormal occurs, the takeoff is to be terminated. Anderson felt the big plane veer slightly left by a few degrees when he pulled back on the control wheel to fly off into the morning darkness. There'd been no reason or call to abort. The landing gear crept off the runway past the edge lights.

He banked the airliner to the right to line up on the remaining runway and then made a sharp left turn calling out: "Gear up." Zeiders responded, "Gear up!"

Even student pilots did not run off runways. Four-thousand-hour airline pilots especially never did. Anderson said to his copilot, "If I had an explanation for that, I'd give it to you!" There'd been no crosswind pushing them to the side of the runway. He knew he'd be in trouble. The thought flashed through his mind for a very brief second, then he was too busy fighting for his life to worry about it.

Trip to Dallas

While the two pilots struggled to save their craft, up in Portland, Alvin F. Oien, Sr.'s family still slept; after all it was still just after 5 a.m. This day was to be the last one before the big trip to Dallas by way of San Francisco in The Viking. There were only minor last minute chores to be done at the trim suburban home. The hotels could get along without his guidance for a week or so. Alvin trusted his employees now that the embezzling manager was fired. In Dallas with his son, he'd concoct a financial strategy to recoup his losses from the thievery.

Over the past days Oien's wife Phyllis, carefully packed the three suitcases, one for her, one for Alvin and one for their daughter, Carla. She made sure Carla's swim gear came along. In Dallas, Alvin's son's house had a swimming pool. Phyllis canceled the mail and newspaper delivery and arranged for the family dog to stay with the neighbors down the street. The Portland Riding Club agreed to care for Terry, Carla's horse. A very organized woman, Phyllis wrote out the combination to the basement safe and placed it inside a letter. Then she hung an extra house key behind a picture in the living room in case anyone needed the key. She really looked forward to some time in warm weather. Though she'd lived in the Northwest since after the war, like most people from the region she welcomed any sort of break in warmer, drier weather, especially at winter's end.

By Friday, Phyllis just needed to drop the family dog off and pick up Carla from Jackson High School later in the afternoon. If all went well, soon they'd all be in Texas visiting with Al's son and daughter-in-law, Sally. Phyllis made sure she packed some of her homemade plum jelly in the suitcases as a gift for Sally. The plums came from the backyard trees.

While the family slept, over the Klamath Falls airport, Anderson found out the hard way how ice destroys the smooth flow of air over the wings, which is so necessary to keep an airplane aloft. Radar records showed rapid oscillations in vertical speed. The airliner did not follow anything like a published departure procedure (DP). These special routes are designed to keep aircraft clear of rising terrain (also known as mountains). Near Klamath Falls, the DP kept aircraft taking off from Runway 14 far from Stukel Mountain. The radar path showed the stricken craft turning directly toward Stukel right after launching.

Anderson called for "flaps up" then asked for radar assistance from the tower. By asking the tower in Klamath Falls if they had radar contact, it meant something out of the ordinary happened. The Fairchild's clearance advised them to contact the departure controllers once airborne.

The CVR recorded Anderson saying:

"Ah, hang on here" and "I'm gonna make a tight turn here."

Then, the alarming shouts:

"I don't know where the hills are!"

The most important rule in flying is to know where the ground lies. Trouble starts when this situational awareness is lost. Sometimes radar vectors help a disoriented pilot to get back on track, or the pilot can figure it out on his own, as long as he has control of the plane but these airliner pilots had no control.

They were in a predicament much like driving a car down a dark interstate with no headlights on, when suddenly the steering wheel falls off. Anderson and Zeider's plane flew off into the dark by itself, the icy wings failing to respond to any control inputs.

"Get us on radar real quick!" one of the pilots shouted. At 0502, Anderson asked the control tower if the plane showed on radar again, then muttered an expletive (often the last thing heard on a CVR when the tape is played back after an accident).

Just four miles from the airport the Fairchild drove nose first into Stukel Mountain, crashing into the ground at over 184 miles per hour. It cartwheeled uphill, ripping into a million pieces, spreading its parts and human remains over a wide debris field at the 2,000-foot mark. Hundreds of gallons of Jet A fuel sprayed from the imploded fuel tanks over the mess that had just been an airliner carrying four people. Then the Jet A exploded and burned.

Just after it took off, six residents had heard the aircraft fly over their houses at a very low height; one reported hearing the impact. Someone reported a flash. There were no survivors.

Twenty minutes after the F-27 departed Klamath Falls Airport, an Air Force operations officer drove a pickup truck onto Runway 14 in the heavy snowfall. All the runway sidelights were illuminated. The wheel tracks in the snow clearly marked the path where flight 720 mowed over the runway lights during the takeoff roll. The wheels left marks 12 feet off the side of the runway. Soon the heavy snow obliterated the evidence. Rescue crews mobilized towards the fire on Stukel Mountain.

As the crews made their way to the burning wreckage of Flight

720, at Jackson High School in Portland, fifteen-year-old Carla Corbus made important arrangements with Mr. Donald Barr, her biology teacher. Carla, a volunteer lab assistant, cared for the lab's tropical fish and fruit fly cultures. She gave detailed instructions written on index cards to Barr, making sure he passed on proper care-taking directions for the aquariums and cultures during her week-long absence. The cultures were important to her. An A student, she decided to become either a chemist or biologist after graduation. Carla explained to the teacher she'd be flying to San Francisco in her stepfather's plane the next morning. From San Francisco they'd take an airliner to Texas to visit Al Jr. and his wife Sally. Carla looked forward to the spring break trip.

By the time aviation investigators visited Stukel Mountain a few days later, the once West Coast airliner remnants lay buried beneath a blanket of white, making it impossible to determine much of anything, but back in the Washington, D.C. National Transportation Safety Board (NTSB) lab, the ghost voices in the Black Box tape solved the mystery of the accident.

As part of an extensive follow up query, other Fairchild F-27 pilots reported the left-turning tendencies when flying with loads of ice. The NTSB ultimately blamed Zeiders and Anderson for the accident. They should have deiced the plane prior to flight. Anderson, as pilot-in-command failed in using his authority.

The meteorological factors that led to the icy conditions were listed in the NTSB report as this: Surface and upper air charts showed a low pressure system centered just off the Washington coast. Cold, moist, unstable air prevailed over southern Oregon and a diffuse quasi-stationary front was located not far south of Klamath Falls. Mixed rain and snow began at Klamath Falls at 1900 on March 9 and remained there until the following day. The report gave specific readings for the area airports, and then cited the general area forecast for southern Oregon as low ceilings and visibilities, mountains obscured, frequent moderate turbulence and icing, strong updrafts and downdrafts east of the mountains. The terminal forecast for Klamath Falls for all the next day matched closely the report given by the control tower to the Fairchild as it made its final taxi.

The Fairchild and her victims were just the start of the death toll; the storm caused other casualties that week. Some in cars, some in boats, some in planes. Some victims of the storm took a longer time to die.

From Portland to San Francisco

By Friday night, when Alvin Oien fueled and loaded the Viking with the suitcases, another cold whip from the system offshore lashed out for the West Coast. As it flailed eastward it inhaled water from the ocean's rough surface. Friday night was a good night not to be at sea. As the storm approached the coast, the scuzzy clouds in the Oregon/California border country dissolved quietly to the southeast. The weather gurus stated that during the weekend the sky would fall once again but maybe the clearing was good enough to squeak through from Portland to San Francisco.

The ice on the wings that snuffed the Fairchild as effectively as a terrorist's bomb originated as seawater from the Pacific Ocean. As the low-pressure vortex spun toward the West Coast across the North Pacific, it vacuumed water up by the ton. A big storm pattern was familiar to people of the west in 1967, already the record books showed off-the-charts precipitation levels for Northern California.

The basic cause of all weather is this: Uneven heating of the earth by the sun. This inconsistent heating changes the temperatures of the overlying air mass. After heating, things get complicated as relentless and chaotic mixing of the air masses follow. This stirring is caused by the spinning earth, the Coriolis effect, and terrain shape, among other factors.

On March 10, the storm that bogged down the west began as a pressure cell from the Aleutian Low, a semi-permanent airmass above the North Pacific. The cell gathered force as it spun easterly, mixing with the warmer air from the southern waters. Butting against the Hawaiian High, the carrier of warm, high pressure air that keeps the southern half of California so temperate, the low pressure juggernaut finally stalled off the Washington Coast. Seeking balance, it regurgitated its moisture onto the land, unloading itself as rain, snow and fog, and the ice and snow that fatally coated the airliner's wings.

Creeping its way over the Coast Range of mountains, the system promised nothing but misery for travelers. Anyone transiting the area risked passing through the gauntlet of snow, sleet, freezing rain, and icy winds that started as a simple meeting of air masses far away. While the National Weather Service issued travel advisories for motorists, the aviation forecast warned generally of the poor flying weather. The forecaster's brew included temporary conditions of "partial clearing." Up in Portland, Alvin Oien banked on getting south through one of the periods of clearing air. Luck usually favored the bold and boldness was his lifetime friend. He knew about Flight 720, it was all over the news and if morning conditions warranted, he'd scrub the mission. If not, soon he'd be in Frisco or Dallas.

Anyone transiting the area risked passing through the gauntlet of snow, sleet, freezing rain, and icy winds

CHAPTER 2

The Boss and His Plane

THE NEXT MORNING, AS the remains of the West Coast airliner lay buried under the fresh snow of Stukel Mountain, the people of greater Portland, Oregon awoke beneath a canopy of gray overcast sky. Chilly mist fell from the gloom as a blue and grey Cessna 195 sporting a large red rose painted on the tail taxied across the Beaverton Airport to the runway. Near the rose in big letters it said: Portland Oregon. A soft roar echoed across the quiet grass airfield as the pilot's strong hand firewalled the throttle knob, pouring the coals to the 300 HP Jacobs radial engine. Crazy about flying, Alvin F. Oien likely grinned as the plane, with her elegant rounded wings and tail, quickly lifted from the green runway and climbed above the city. Then it turned south, paralleling the new interstate toward Salem. Alvin really was crazy about flying.

The people in the plane had big plans. In a few hours they'd land in California. The takeoff marked their vacation to warmer climates. Few people were at the airport during this early hour. Beaverton Airport, actually an old farm field, catered to private pilots and the rainy skies did not beckon private pilots to airfields the way a clear and sunny weekend did. One man on the ground recalled the plane's takeoff and knew all about the flight. He owned a business there called Rupert's Flying Service.

Walt Rupert, a pioneer northwest aviator, sold a load of aviation gas to the pilot of the gray-and-blue Cessna the previous day. Nineteen dollars of 80-octane Avgas topped the fuel tanks. Rupert and the pilot were longtime flying friends and Rupert knew the plane intimately, having signed off her annual inspections since the late 1950s. Standing at the fuel pumps that morning, he'd watched and waved to the family in the plane as it taxied past and disappeared off to the south. Although he did not know it while he waved, he'd

19

never turn another wrench on that airplane or see its passengers again. Rupert never imagined the hell his friends were to endure.

In the pilot seat of the rose-tailed plane sat the strong-willed Oien Sr. His family called him the "Boss." He named his plane The Viking—Al considered himself a Viking too; he was the son of Norwegians. Successful and self-made, he'd battled a path through life like a modern day warrior. He fought his way upwards from the Northwest logging camps since the age of eight, and all along the way dominated his employees and family. The captions of his pictures in the family album simply said "The Boss," and he looked like someone used to being in charge.

At fifty-nine-years old, the bull-necked Oien owned a class B hotel property in east downtown Portland called the Clifford, which sat at the west approach to the Morrison Street Bridge. Proudly, the pen and wallet he carried in his pockets advertised the Clifford. He'd owned the property since the war along with a few apartment buildings as well as another hotel in Baker.

A hardass to everyone around him, those who knew him best said that, to be fair, he was tougher on himself than on anyone else. His parents were South Dakota ranchers who brought young Alvin to the ranch in Buckeye, North Dakota after his birth in Minnesota in 1908.

Following the funeral of his father, relatives from Washington invited Oien on the train back west, letting him carry all he owned, as long as it fit in a shoebox. In the Washington woods, he worked as a whistle punk and cook's helper for a logging crew in Aberdeen. Through the harsh school of the woods, he made his way into the world of men by transitioning from cook's helper to whistle punk to choker setter.

Seeking a better occupation, in the 1920s in Portland he attended Adcox Aviation School while washing dishes at Manning's Restaurant downtown. At Manning's he met Laura, a farm girl from nearby Sauvie's Island on the Columbia River. Laura became his first wife, and in time she gave birth to three sons and one daughter.

Al served in the Oregon National Guard, worked as a self-employed trucker, owned a fuel delivery business, and then bought the Clifford Hotel on Morrison Street in downtown Portland after the war ended in 1945. After that he learned to fly. Not a bad result for a man who started out with very little.

People knew him for his blunt ways. He spoke what he believed, no matter the cost. When he interviewed truck drivers for his company, he'd test them by letting them take a spin around the Portland streets. He'd warn them not to drive too far because if he did not like the way they drove, he'd kick them out and they'd have to walk back.

Over the years he'd proven himself to be someone who lived by his own laws. After the war, he bought a new Luscombe Sedan airplane and wrecked it in the fog near Chehalis, Washington. He hurt his back in the crash and doctors hospitalized him in a body cast. The cast stayed on until he found out the Oregon Pilot Association planned a flying goodwill tour to Cuba. Medical advice be damned for Alvin F. Oien, Sr.—Cuba called and he needed to fly, so he cut the cast off, bought a Fairchild 24 to replace the wrecked Luscombe, and flew to Cuba and back. Later he piloted the Fairchild all over North America, amassing thousands of flight hours. In the late fifties he replaced the Fairchild 24 with The Viking.

Always confident, his eldest son and namesake remembered a time in the '40s when Al Sr. learned to fly at Hillsboro Airport in a Stinson Voyager. On that day, Junior sat in the backseat with a friend while his father flew from the pilot seat. The small plane's engine quit abruptly while at altitude. The instructor gasped while Al Sr. confidently juggled the engine controls, finding the sweet spot that kept the engine going as the plane limped back to the airfield. Senior never blew his cool. He just sat in that pilot's seat smiling and singing his "di-dee-do" song as if he'd not a care in the world. He constantly belted out the di-dee-do song, a wordless tune sung in a gravelly baritone.

His history shows some well-publicized scrapes with the law, splayed out on the pages of the Oregonian, the oldest continuously published newspaper on the West Coast. These sorts of legal predicaments would not even register in the news of today's Portland crimes. Back then, without daily carnage to report, when a man stood up for himself and his natural rights, he stood out to reporters, especially if he owned a downtown property.

He once fought the cops over a traffic stop and lost the ensuing lawsuit for false arrest. Then a fight with hotel tenants over their slow payment of rent used more of the Oregonian's ink. Once, Al landed without runway lights at Vancouver, just across the river from Portland, and lost his flying privileges in Washington State for a short time. His cousin, Edna Hallet, said he never abided by the speed limits and made his own traffic rules. All the city traffic cops knew him, and when he got stopped he'd just snap his fingers and tell the officers to hurry and write the citation. He needed to get to work so his tax dollars could pay their wages.

At the Clifford, he once climbed down a ladder while replacing a fixture and flattened a hotel union organizer who'd dared tell him he could not do maintenance on his own hotel. No one told the Old Man what to do ever, whether union organizer or doctor, wife or police officer, and he seemed unafraid to pay the consequences.

Like a real warrior, the Old Man fought certain things, but inside he carried a good heart. He belonged to several social clubs. He never missed a meeting of the Knight of Pythias or the Optimists Club. In the summer he'd fly over the Optimist Club camps in the Oregon hills in his Fairchild or Cessna and dump buckets of candy down to the needy kids.

Years later, nephew Steve Hausheer of Illinois, recalled the Old Man: In the late 1950s, at age four, Steve was confronted by his mountainous Uncle over his inability to tie his own shoes. Al Sr., a larger-than-life figure who flew planes and was married to Aunt Phyllis, shouted down to him, "You are four years old and you don't know how to tie your own shoes!" This uncle terrified him at first, but then took him aside and carefully taught him how to do the task. That, however, was not before Al Sr. explained that he'd dunk Steve in the toilet if he failed to learn. To Steve, this big man of the west became an instant hero.

Al Sr. and his second wife Phyllis flew their airplane all over the country and were enthusiastic members of the Oregon Pilots Association (OPA). He became the projectionist at the local chapter and ran the movie projector while Phyllis served as the food coordinator. Phyllis was even elected club treasurer for 1964. They participated in Northwest fly-ins and aerial search and rescues when planes went missing. One year, the Old Man had even located a crashed plane and the downed survivors.

Alvin Oien, Sr. grew up in America at a time when aviation was promoted as the new gospel, and in the late 1920s, flying was indeed a miracle. The hero fighter pilots of World War One, the barnstormers of the '20s, and mail pilots like Jack Knight and Charles

Lindbergh stirred the imaginations of millions of air-minded youths. Man now had the ability to travel to far off lands and places. To a red-blooded young man of that time, airplanes offered unlimited excitement, romance, and adventure, not to mention prestige. Airplanes made Alvin

One year, the Old Man had even located a crashed plane and the downed survivors.

Oien crazy. Over the years, he owned several small planes and flew as much as possible. Bred from the old school of flying, he'd point the nose of his plane and just fly away, simply heading where he wanted to go.

Following fifty-one years of hard work, the Boss established himself as a businessman, family man, and a competent pilot of small planes. The foul weather on March 11, 1967 over the wild mountain country of Southern Oregon and Northern California could challenge any pilot, but if anyone could fly a plane from Portland, Oregon to San Francisco, California it was Alvin F. Oien, Sr. He knew the route, and his trusty Cessna 195 carrying U.S. registration number N9388A, completely suited the flight.

The Viking

The best he ever owned and a mountain goat of a plane, The Viking could do anything the Boss asked, and it could go just about anywhere. He flew her for over one hundred hours a year tending to his hotel properties and family. Back in Bullock, South Dakota he'd land his plane at the family ranch run by his brother Percy. The small boy who left on the train for the West Coast with nothing ended up owning a hotel and flying back to the ranch in his airplanes to say hello, which always earned a note in the local paper, the *Buffalo Times-Herald*. In the old family album you find pictures of the red-nosed Cessna parked at Percy's ranch, Oien having landed her skillfully on the sod and taxied through the mud to the small house where his mother still lived.

Like its owner, The Viking was strong, capable and reliable. A Cadillac of aircraft back when Cadillacs set the standard for quality, Cessna made over a thousand of the 195s. Manufactured in Wichita, Kansas from the 1940s until the mid-1950s, The Viking seemed out of date even then. Its design harkened back to pre-war times with the round seven-cylindered Jacobs 300 HP engine.

The 195 seated five people in comfort as well as their baggage. It flew in excess of 150 miles per hour, making it a very efficient plane for a businessman to operate during a time without the widespread air service of today. Some commuter airlines flew them as Businessliners. Flying from Canada to Mexico in a Cessna 195 could be done in one long day. Coast to coast could be flown in two days. Al purchased the nine-year-old plane in 1958.

Starting out in 1949, fresh off the Cessna factory line after being tested and shaken down by famed test pilot Mort Brown, The Viking carried her N-number designation in large blue letters on her gray sides. Owned and flown as a business transport for Northern Natural Gas of Omaha, she ended up with Oien after being flown for a few thousand hours throughout the Midwest. Over the 1950s, equipment was added to expand the aircraft's already admirable capabilities, making it an almost go-anywhere-anytime machine. Upgrading from the smaller Fairchild 24 to the Cessna allowed Al to fly even more. The logbooks for N9388A show he flew her throughout the West and to the Midwest. Her yearly inspections were signed off by Walter Rupert, the man who witnessed the plane's last takeoff.

In 1967, Cessna 195s still out-performed their modern counterparts. Even today, the Cessna 195 is a beautiful, high performance, sought-after classic. Those who fly them say it is hard to be humble when you fly a Cessna 195.

Rated by the government as a VFR (visual flight rules) only private pilot, though his Cessna carried instruments for "blind flying" in clouds, Al flew with weather restrictions. There are two types of flying: visual flight and flight by reference to instruments. Being a visually rated pilot forced him to fly only when he could see the ground. VFR pilots like him operated with mandates letting them operate only when the cloud ceiling stayed one thousand feet above the ground and the visibility remained three miles or better. On March 11, 1967, as long as Al saw three miles and the clouds floated at least a thousand feet above the ground, he could fly his plane south from Oregon to California.

All along the new I-5 superhighway corridor, which connected the northwest with the southwest, the airports reported improving conditions. No reason existed to go back to the Beaverton Airport once The Viking's people returned Walter Rupert's wave goodbye and took off. The weather looked to improve as the small plane flew toward California.

The weather charts for the early morning of March 11 showed some cloud cover along the flight path, but by morning, when The Viking sped toward San Francisco, the forecaster predicted a short window of partial clearing, this period being the gap seen as a boxcar speeds past. If a person can jump through the gap they don't get slammed. If they misjudge, they don't survive.

If The Viking could get to California during the clear period, the plane would easily arrive in San Francisco by mid-afternoon. If Al needed to climb the plane high to cross some weather, he'd carefully do so. The craggy peaks of the western mountains held fast onto the messy line of clouds in the frontal passage, obscuring peaks, filling valleys and passes, and clogging the roads with snow, rain, and fog.

Climbing above turbulent weather has always been a strategy to avoid poor flying conditions below. Going high across a band of weather then descending into clear air is the aviation version of vaulting a muddy puddle to keep one's shoes clean. If necessary, The Viking would carry them over the weather along their route and they'd descend down into the long Sacramento Valley in good visibility and then work their way to San Francisco by midday.

Although conditions for flight were not ideal the day The Viking launched, to a gambler betting on the odds of a small plane getting through from Oregon to California on March 11, 1967, Al F. Oien and his trusty Viking were a reasonable bet. Al had plenty of experience, flew a good plane, and he knew the ins and outs of the route.

CHAPTER 3

The Family

I N ADDITION TO THE Old Man, The Viking carried Al's second wife Phyllis and their daughter, Carla Corbus. A big woman, dark-haired and pretty, Phyllis sat in the right front copilot's seat while her daughter sat in the big bench seat in the back of the plane. Carla still used her biological father's surname, Corbus, although she didn't remember him and considered Alvin her only father. But if Oien Sr. had legally adopted the girl, her survivor benefits from the death of William Corbus would stop. Some news stories reported Carla as a reluctant passenger; other accounts stated she was happy to be going along on a trip over spring break.

They planned to make an airline connection in San Francisco and fly to Dallas where they'd enjoy a vacation in the warm Texas sun while visiting with Al Jr. and his wife Sally. It was good having an airline pilot in the family. The Delta Airline connection in San Francisco came compliments of Junior, who'd provided discounted tickets to his father's small family. As if to head off any trouble from the Old Man, Al Jr. carefully explained in a letter that because the family was traveling standby they may be bumped for someone holding a higher priority, and that Al Sr. must behave himself if he found himself waiting for another flight. "Just remember on a pass you are the bottom man on the totem pole and they take a dim view of employees relatives giving them any static if they have to bump you for a full fare or military half fare or if they do not have enough meals to go around," the letter read.

Phyllis

Carla's mother, Phyllis Hausheer-Oien, hailed from Chicago, Illinois. The daughter of Walter and Martha Hausheer, she'd been born just before Christmas at the Grant Hospital on December 21,

1922. Walter, of Swiss descent, worked as an importer of high quality medical equipment. An intellectual family, the Hausheers instilled in their daughter a love for nature, science, philosophy, music, and education. An accomplished pianist, stargazer, birdwatcher, horse rider, and outdoors woman, Phyllis loved all things of the natural world. After high school, she studied general education at Michigan State University, then graduated from Northwestern University in Chicago with a bachelor's degree in nursing, just after the end of World War II. During the war she organized scrap drives. Photos from the Chicago newspapers show her rallying people to donate metal for the war effort. In 1944 she joined the Army Nursing Corps. Her grade transcripts from the university show her as a far-above average student with plenty of A and B grades. After graduation she became a nurse at the Evanston Hospital. Described as rather "severe" in manner but very compassionate to patients, she ran the children's polio ward in her organized, methodical way.

She met William Corbus, a war veteran, sometime in the late '40s in Chicago. Corbus, the son of well-known physician Dr. Budd Clarke Corbus, served in the U.S. Army First Engineer's Combat Battalion, Company A, and saw heavy fighting in the European Theatre of Operations during WWII. The First Engineers were the most decorated combat engineer battalion in the U.S. Army, seeing combat in the African desert as well as the Battle of the Bulge. Corbus' stellar war record did not help him when he returned home. Bill Corbus did not integrate well into civilian life and he returned from the war a wreck. He embarrassed his family by abusing drugs and alcohol, according to Alvin's cousin, Edna Franett.

Phyllis and Corbus married in 1950. Within one year Phyllis became pregnant with Carla. In Chicago, while Phyllis worked and cared for her young daughter, William Corbus visited drugstores, seeking medication to eradicate his mental pain. It is likely his affliction would later be labeled PTSD, an undiscovered malady in the 1940s when veterans were said to suffer from "combat fatigue."

Not willing to give up easily, Phyllis and William thought a change of venue could revive his health and their failing marriage. Perhaps a better climate and change of scenery would be enough to turn things around. A fresh start away from family on the booming West Coast appealed to the couple. Phyllis applied for Registered Nursing Certificates in California, Oregon, and Washington. Seattle was their choice for the family's new life. The city boomed with the expansion of the Boeing Company. The aircraft manufacturer could not make enough jet bombers for the U.S. Air Force. The couple moved to the wet city where Phyllis took a job at the Ballard hospital and they rented a small apartment. After a very short time, the fresh start turned sour. Typical of addicts, William Corbus continued his old ways, visiting drugstore after drugstore seeking medication.

Carla

Through her medical training, no-nonsense Phyllis knew enough about the hopelessness of addiction and sent her husband packing back to his Chicago family. She filed for a divorce which soon was granted and began the difficult—and uncommon for the time—task of raising her young daughter as a single working parent. In 1955, William Corbus killed himself at a veteran's hospital in Chicago. After that, for the rest of Carla's short life, Mrs. Budd Clarke Corbus provided funds so her granddaughter could grow up with nice things. Over her lifetime, Carla studied ballet, piano, and horsemanship, most paid for by her doting grandmother. Carla never remembered her real father and grew up with Al Sr. as her dad.

At fifteen, far from the giggling, adoring daughter who once sat on Alvin, Sr.'s knee while he sang the "di-dee-do" song to her, Carla was an attractive, intelligent teen with long dark hair, blue eyes, and a pretty face. At Jackson High School she played piano, sang in the choir, carried excellent grades, and worked as an assistant in the biology lab. After high school she planned to study chemistry or biology, having inherited a love for science from Phyllis. In Donald Barr's biology lab at Jackson High, she cared for twelve tanks of tropical fish on her own time as a volunteer assistant. Over the years she'd become an accomplished horse rider as a member of the Portland Riding Club, riding her own horse named Terry. She covered her bedroom wall with ribbons she'd accumulated from riding events since she was a young child.

Maybe the headlines spooked him. The Old Man must not have noticed during the youth rebellion of the late 1960s that his step-daughter played no role other than being a responsible, admirable teenager. Nothing in her life suggested she chose anything but a mature path. Over the previous year, Carla had fallen in love with a clean-cut young man a few years older than her named Rick. Without good reason he'd been barred from seeing her by the Old Man and considered himself persona non grata at the Oien home in Southwest Portland. For that he railed against the Old Man, complaining of his "Victorian" ways.

Fashion restrictions loosened for Carla from 1966 into 1967. She'd been allowed to grow her hair longer in the preferred style of the time, though she wore what classmates described as a "uniform" of conservative sweaters and skirts when other girls began wearing short dresses and jeans to school. While her parents, particularly the Old Man, may have fretted over her, Carla's teachers thought of her as nothing but a responsible, level-headed young woman with a very bright future.

Before the Old Man took over their lives, Carla and Phyllis lived alone in Seattle in a rented two-story apartment not far from the Ballard Hospital. On weekends the two explored the Puget Sound beaches and the coastal mountains. Often Phyllis drove her Pontiac into the Cascade Mountains just east of Seattle where she camped with Carla in a small tent. Other times they ice-skated at the Highland Ice Arena in North Seattle. Carla showed great promise as a skater. Phyllis bought her costumes and arranged for skating lessons. Phyllis herself took up the sport just to encourage her fatherless daughter.

On the lower floor of the apartment complex in Ballard lived a cousin of the Old Man named Edna Hallett, another Midwestern transplant to Seattle. She'd migrated to the West Coast with her husband after the war, searching for better opportunities. Over the war years she'd worked together with the Old Man's sister at a Minnesota munitions plant.

At first the upper and lower floor neighbors had disputes over noise, but soon Phyllis and Edna Hallet buried the hatchet and cultivated a friendship that lasted for the rest of Phyllis's life. As it deepened, their two children became friends too. Carla and Kelly, Edna's son, also happened to be the same age. At ninety-five years young, Edna still laughs about the occasion when she stopped Carla and Kelly from jumping out the second floor windows. The two kids pinned towels on their backs like Superman and thought if they ran fast enough toward the open window, they'd fly. Carla and Kelly stayed close, though they lived in separate cities. Edna saw Carla last on New Year's Day in 1967 when she was picked up from her home by Al and Phyllis. She watched Carla drive off with her parents and never heard from any of them again, something she never expected.

In Ballard, the two small families planned their weekend excursions together, traveling through the Northwest and splitting costs. During the weekdays, Edna watched out for Carla while Phyllis worked. Phyllis returned the favor by helping Edna clean house on weekends. Sometimes Edna, Phyllis, and the kids drove all the way south to Portland, where they skied on the slopes of Mount Hood or played on the Oregon beaches. The Old Man allowed Cousin Edna and any of her friends to stay at his house in Portland.

Al always loved to fly and took people for rides. A few times he took Phyllis and Edna flying over the Oregon countryside in the Fairchild 24 or the Cessna 195. Edna said he loved turning the controls over to whoever sat next to him, whether they wanted to fly or not. He'd done so for her a few times and it scared her silly to steer the plane. Pilot or not, whoever got the controls flew while Al sang the di-dee-do song and laughed at their efforts. Edna hated the Old Man's casual ways with the plane, but Phyllis loved steering the aircraft on her own. She knew he wouldn't let anything happen. Afterward she'd exclaim about Al Sr., "He's the kind of guy I like!" Al, who suffered his own marital demons, provided a bright spot in Phyllis's lonely life.

Al showed up in Ballard to visit Edna and, by default, to see Phyllis. Sometimes he just happened to show up on weekends when the two ladies had planned an event. Edna recalled a time in the Cascades when Al went along to the high country with the two women to show them how to apply tire chains. While working outside on the tires, Al and Phyllis wrestled in the snow. Al, a fireplug of a man and "monstrously strong," completely lifted Phyllis over his head and then placed her upside down in a snow bank. Al's brute strength impressed Phyllis. Again, she commented to Edna, "That's the kind of guy I like!" More and more, Edna noticed the remark. Phyllis seemed to glow around Al. Edna was aware of the mutual attraction between her cousin and her friend. Al's good looks did not help the situation.

In time he attested to his marital problems and afterwards began a discreet relationship with Phyllis. On weekends he traveled to Seattle and furtively visited Phyllis. Edna, who also worked at the hospital as a receptionist, heard of Phyllis's weekends with Al through the hospital gossip grapevine. A marriage proposal resulted, to the dismay of Edna, who said she "cried and cried" for two days when she found out her friend and cousin planned to marry. She felt responsible for the breakup of Al's marriage and the potential ruination of Phyllis's life. Edna knew Al acted like a monster to his first wife Laura, and felt Phyllis had doomed herself to a life of terrible abuse and servitude.

Edna underestimated her friend's tenacity. To many people the couple seemed an incredible mismatch: Phyllis, with her education, musical skills, scientific interests and her young daughter, and Al Sr., the gruff, tough, ex-logger who'd come up the hard way. Phyllis's father wrote about Al's "Cessna-consciousness" and blessed their union, stating Phyllis knew what was best for her. In the end though, Phyllis summed up her years with Al in a final letter about her second husband: "[R]eally Al has been very good to us," and he was.

Phyllis, a strong, competent, confident woman herself, made a great counterpart to the senior Oien. Phyllis amazed Al Jr. after meeting him for the first time over dinner at a Portland restaurant.

Phyllis, a strong, competent, confident woman herself, made a great counterpart to the senior Oien.

Around Phyllis, Al Sr. did no fuming, blustering, threatening, or swearing. Later she explained to the amazed son that the first time Senior misbehaved at home, she'd just pointed to the door and said, "Out!" Hanging his head, the chagrined man went outside in the rain and sat on the curb until he agreed to mind his manners—only then could he return to the shelter of the house. And so the couple got along well enough. Sometimes in front of company Al slapped Phyllis on the bottom and loudly exclaimed, "Now there's a lot to love there!"

Life in Portland

For the first time in a long time, Phyllis's and Carla's life moved forward with the financial and domestic stability Al brought to her life. They set up a small home in Southeast Portland and went about raising Carla. The home glistened to the hospital-clean standards of Phyllis's professional training and personal taste. Even the copper-bottomed pots in the kitchen shone like mirrors. During the day Phyllis maintained the modest house, taking careful inventory of its contents. One thing remaining more than forty years later is a bird list of twenty-four avian sightings. Carefully, on the back of a checking account slip associated with an apartment property of Al's, she documented the common Northwest songbirds. On another list Phyllis even included a junk handle from a broken mug. After she was gone, people found she'd listed almost all the family possessions on meticulous lists, the same way the contents of the suitcases packed for the Texas vacation were itemized.

By 1967, Carla attended Jackson High School, and each day Al drove his International truck downtown to the Clifford Hotel where he worked in his hole-in-the-wall office, the walls covered with humorous cartoons, one saying, "For a good Fourth, don't buy a fifth on the third," as well as aircraft photos and clippings. He also had an affinity for the strange pictures of people with double-deck-ered eyes that were popular at the time. 1967 was a memorable year in America. The Vietnam War was in full swing, the youth were rebelling or so it seemed, the Summer of Love was to come and the assassinations of King and Kennedy were still a year away.

Alvin Oien became the only father Carla ever really remembered. The rugged ex-logger would tenderly throw the giggling girl on his knee and belt out his trademark di-dee-do song in his loud, deep voice. Over their time together he taught her to love the outdoors, fishing, and photography. Phyllis taught her daughter to love garden-ing, books, animals, and science.

Alvin Oien became the only father Carla ever really remembered.

By the day of the flight, trouble was looming due to Carla and Rick's relationship. The two parents thought a trip south during spring break could help keep her from getting in too deep with the boy. But Rick and Carla planned to marry when they were both of legal age. They'd announce the plans on Carla's upcoming sixteenth birthday. As a surprise gift, Rick had paid for his mother's engagement ring to be reset for the special day.

Having struggled with her first marriage, Phyllis held no illusions about immature romance. Al knew the pain of a troubled marriage too. People said the Old Man and his first wife Laura never exchanged a civil word after the death of their young daughter in the late 1930s, after she succumbed to a communicable childhood disease. No doubt the parents felt they were protecting their daughter by keeping Carla away from an early love affair.

Adding more ingredients to the troublesome brew, there'd been some business difficulties with the Baker Motor Inn, Oien's hotel property at Baker, Oregon. Finding himself deeply in debt from a thieving manager, Al sought advice from his eldest son, who'd received a formal education. The junior Oien suggested to his father that he borrow as much money as possible against the Clifford Hotel—the more the better, then no bank would dare foreclose on a hotel used as collateral. The die was cast for a remedy. The plan intrigued the Old Man, so the Texas trip came about with the younger Oien providing the discounted Delta Airline tickets and an offer for everyone to stay at his Dallas house. "Make sure to pack swimsuits!" he wrote.

Since Delta did not fly into Portland, flying the comfortable old Viking to San Francisco would get them to the closest Delta gate. The composite trip fulfilled many things: Carla would be away from Rick for a week; the Old Man loved to fly; there was the promise of a financial remedy to his hotel woes; along with a family vacation to warmer climates. Phyllis, after a gloomy Northwest winter, looked forward to some warm weather. In her typical, methodical way, she prepared the house for the family's absence. She put the dog in the care of a neighbor, packed and inventoried the suitcases, and even freshly shined the copper-bottomed pots hanging in the kitchen. She hid a house key behind a picture frame in case anything happened on the trip.

Carla told Mr. Barr in the biology lab she looked forward to some warm weather, but her boyfriend Rick claimed she would've preferred staying in Portland, where they could have spent the week riding horses and relaxing together. In only a little over a month they planned to announce their engagement.

The Oien Boys

Al Jr. made a life for himself when, at the age of fifteen, he moved out of the family home, saying enough was enough. Back then he could not stand his father because his parents fought constantly. On top of that, Al Sr. constantly belittled his three sons, Al, Chuck, and Ron, claiming they'd never amount to much. The younger boys, Ron and Chuck were stuck with the Old Man being too young to escape home. The boys learned to stay out of the Old Man's way, avoiding him like the plague. Junior believed his parents never came to terms with the 1937 loss of his sister. She'd been sick and the doctors had stated if she lived past her sixth birthday, she'd survive. She died just after her sixth birthday. Junior graduated at the top of his high school class, paying his own way for the last two years.

When his eldest son asked for college tuition, Al Sr. just laughed at him and told him to "go to hell," that he was not a bank. He recommended his son just earn the tuition himself, so Junior worked his way through Reed College in Oregon pumping gas at night and studying during the day. In the mid-1950s he transferred from the Oregon National Guard and entered flight training. He became a United States Air Force pilot and flew everything. His USAF record shows he flew the last B-17s, the C-123, the DC-4, the C-131, the SA-16 as well as fighters and helicopters. A missile chaser pilot at Cape Canaveral for much of his active duty, he married an Air Force nurse named Sally.

In the early 1960s, Al Jr. took a job flying C-46s for Delta Airlines. By 1967 he was co-piloting DC jets to the west coast from the Dallas base. By becoming a military pilot, then airline pilot, in the world of aviators, Junior had reached the top of the pyramid, something his hard-charging father neither could nor would ever achieve. Al Jr. was the real deal, as far as flying went. He could fly any airplane, any time, just about anywhere, and he proved it by passing training in the most elite flight school in the world. Al Jr. rose on his own in the world of men, mostly through his own efforts. He became the sort of person any father would be proud to claim as a son. Senior, like many tough-minded fathers, never told his son how proud of him he'd become, though he made sure he told others how his eldest was a "chip off the old block."

Junior's two brothers, Ron and Chuck, had a tough time growing up under the iron rule of their strict Norseman of a father. Being younger, both were forced to live under his wrath longer than Al Jr. who had left home at age 15. The Old Man just could not be civil to any of boys and through their childhoods the trio "tried to stay out of his way, avoiding him like the plague." If the boys were quietly watching TV, to Al Sr. they were displaying their "ineptitude," their laziness…TV watching was just proof-positive of a bleak future.

Around the time of The Viking crash, Ron Oien had not laid eyes on his father since a 1950s incident when he'd been sweeping the hotel basement. Finishing the job, he explained he needed to leave to attend a Glee Club practice at school. Finding a small patch of unswept floor, Al Sr. became angry and demanded Ron redo the entire job.

This conversation led to a fight and Ron last saw his father while dodging his punches: the young boy fled from the basement and onto a sidewalk and he never looked back. By 1967, Ron was employed at Sears Roebuck and doing well. He knew his father checked up on him through mutual friends, but the two never spoke again after the hotel basement incident.

Chuck Oien was the youngest of the trio. In 1967 he as well had not seen or spoken to his father in ten years, except for chance sightings of him around Portland. At that time Chuck was 24 years old and worked nights at the First National Bank, proof-processing checks. He was a member of the Oregon National Guard. Chuck also knew his father checked in on the brothers' activities, but no olive branch of peace was handed from either direction.

After learning a lesson on saving money as a youth, Junior, a cautious man, always tucked away some of his earnings. After his Air Force time he opened some businesses. Between the ventures and his education, he'd developed some skills of his own. In contrast, even in 1967 Al Sr. still kept the Clifford Hotel's financial records in his fedora, and after owning the hotel for over twenty years, had no estate planned. Junior's idea for straightening out the financial mess caused by the thief appealed to Senior.

Now that the hotel business was on the line, Al Sr. would get serious with his affairs and complete his will. The year before he'd looked into it with his attorney and he'd never followed up on the initial paperwork. Like most self-made men, he probably thought he'd never die.

There'd been a long period of estrangement between the father and his eldest son throughout the 1950s up until 1960. But as time passed, the son now entering his early thirties could see things from a different perspective. Perhaps his father had some justification for divorcing Laura. The upcoming visit between the two men signified the growth of a real friendship between two adults, colleagues who could solve troubles together. Al Sr. and daughter-in-law Sally Oien got along well, now, too. She was like a lost daughter to the Senior Oien, stepping in for the little girl who had died in 1937 (who was also called Sally), and Al Sr. became the father she missed.

But the younger sons had not reached their eldest brother's understanding of their father. They'd not reconciled with Al Senior.

The night before The Viking left Oregon, Al Jr. asked Phyllis over the telephone not to say a word about him meeting their flight at the San Francisco airport. He happened to be flying a Delta jet to the West Coast the next day. He'd surprise the Old Man on the ramp at Butler Aviation when The Viking taxied up at 3 p.m. He really looked forward to seeing his father, hosting the family, and helping resolve the crisis.

CHAPTER 4

The Flight

I T WAS A DREARY day when The Viking set out from the Beaverton Airport. Weather reports and forecast records show the Portland air as cloudy but flyable on the morning of March 11. The northern segment of the track lay under an overcast lid of rain clouds, while the southern portion of the route lay under clearer, drier air. The border zone of southern Oregon, where the Fairchild airliner crashed the day before, was like a different planet. The icy air along with the messy, wet cold front had dissipated over Nevada. Being a VFR rated pilot, Al depended on several miles of visibility and a high ceiling to get past the rugged mountains of the Oregon-California border, and each airport reported VFR weather.

Another blast of North Pacific moisture promised to soak the entire West Coast by late Saturday afternoon. The morning showed a window in the wave of explosive impulses peeling away from the offshore low. To the northwest of San Francisco, messy wet clouds still clung to the mountain peaks like fur on a cat brush and lay directly across The Viking's flight path. To get south, Al planned to duck and weave around the clouds like a prizefighter dodging blows. He only needed a few select hours to squeeze past the high country of the Oregon-California border. Midday looked like the ideal time to fly; after that they'd never slip past in The Viking. Saturday was the day, the only day, to fly.

The flight path from Beaverton roughly paralleled the I-5 Interstate Highway corridor south through the farmlands of the lush Willamette River Valley. For most of the route, if there was not an airport a few minutes away, a highway below the wings provided the security of an emergency landing spot. The verdant farm fields gave way to grass covered ranch country with rolling hills, then to rugged pine-covered mountains. Few people lived among these peaks. From Medford, the airway vaulted across the

Siskiyou Mountains to Fort Jones, where it jogged southeast toward Redding. From Redding, if conditions warranted, the plane could fly on directly to San Francisco. Al planned a tentative gas stop in Red Bluff. Stopping there in the early afternoon afforded them a chance to stretch their legs and take a break after being cooped up for a few hours.

From the start, the plane's wing tanks carried enough gas to allow plenty of options. Walt Rupert had seen them filled to the brim. Fuel gives flexibility. For most pilots, the only time an aircraft carries too much is when it is on fire.

In the aftermath of Oien's flight, FAA investigators found at least two pilots who departed Portland for San Francisco in small planes on March 11. One pilot testified he returned to Portland when weather along his route blocked his flight path. The second pilot diverted far to the east of the Sierra Nevada Range to avoid the cloud buildups lying over the craggy mountains of Northern California. Comparing pilots and planes on the same route can only give a rough idea of the flying on a particular route or day.

Conditions change constantly and the view from the cockpit differs each minute. One can only speculate after the fact about a pilot's decision to pass through a region. Weather that scares one pilot can be child's play to another. Experience, equipment, and intestinal fortitude all go into aerial decision making.

For some unknown reason Al left the survival pack behind in the hangar.

What's known for sure is when the little gray and blue plane departed the Beaverton Airport just after 0930 and Walter Rupert watched it take off for the last time, for some unknown reason Al left the survival pack behind in the hangar. Like many pilots, The Boss usually carried extra gear including food, medical supplies, matches, and a rifle as standard equipment on all his flights.

Maybe he thought this: Who really needed survival gear anyway? No law required it, plus the route followed the heavily traveled VOR airways, it paralleled the highways, and there were plenty of other aircraft in the air to call if trouble arose. For most of the flight there were plenty of places to land. No one would be alone out there, except maybe for the leg across the mountains from Medford, Oregon to Red Bluff, California, the area partially obscured under the messy wake of the stationary front that passed. If anything went sour, they could easily get on the ground somewhere or turn around.

For the March 11 flight, the plane's interior was packed with the inventoried suitcases, a five-gallon oil can, and Al's tools. A mechanic at heart since learning the trade at the Adcox Aviation School, Al did a lot of the work on The Viking himself. For field repairs the tools came in handy. Maybe The Viking ran out of room, or its weight with three people, full fuel, bags, and tools exceeded aircraft load limits.

Aircraft performance parameters are based on weight. Maneuverability, speed and controllability are critically affected by weight. Woe to the pilot who overloads his plane. Besides the tools and suitcases, on board were Carla and Phyllis, along with some extra blankets and sweaters. These extra items proved vital.

Not long after departing, over the Newberg VOR at 10 a.m. and at a thousand feet, Al transmitted The Viking's departure time from Beaverton to the Portland Flight Service Station. He filed a visual flight plan for San Francisco, with a possible stop at Red Bluff for fuel. He estimated the time en route to San Francisco at five hours; the fuel endurance of the plane being six hours. A flight plan let someone on the ground know about his plane's route. If The Viking failed to show in San Francisco at the estimated time of arrival, the Flight Service Station operators could begin searching by backtracking to any flight service station or airport along the planned route and determine if there'd been any communication with the missing plane. Often, overdue aircraft were found tied down at airports when pilots changed plans due to weather or mechanical problems, then forgot to relay their change in location to the FSS. Those were the circumstances from Alvin's unneeded search and rescue incident years before, when the La Grande Flight Service Station failed to pass along his changed flight plan.

A search is launched immediately if a pilot overflies a flight plan. Filing the plan made sense and Al's listed the plane, the route, the pilot, fuel endurance, equipment, and number of people on board, along with a contact phone number in the event the aircraft failed to arrive on time. If the plane went down and if Al stuck to the flight as planned, they'd be easily found and rescued. Searchers simply needed to retrace an overdue plane's route in order to find it. Flight Service Stations are ground stations. An FSS usually does not control or direct aircraft traffic. These stations are lifelines to the pilots plying the skies, and are set up throughout the United States—they provide weather and information to pilots.

Early in the flight, the Boss thought ahead and planned for contingencies. Over rainy Eugene, Oregon, he again called Flight Service and checked the conditions ahead for San Francisco, Red Bluff, and Medford, and for winds aloft at 12,000 feet. Flying high in the Cessna would be much preferable to fighting their way down low through cloudy mountain passes.

Up high a pilot can cross bands of weather and terrain and get a bird's eye view of the route, making better time and saving fuel. The Viking could fly efficiently up to 18,000 feet and carried instruments for flying in clouds, though Al planned to operate under the visual flight rules that mandated seeing the ground. Like any trained pilot, he knew how to use the basic instruments to get out of a bind. He'd probably done so more than a time or two over the years in his thousands of flight hours.

The Eugene flight service specialist announced favorable conditions ahead for the plane. Red Bluff carried reports of the lowest weather, with a broken layer of clouds at 2,000 feet with ten miles visibility underneath, plenty for a VFR pilot. The plane bored onward through the wet Oregon air.

Maybe Al sang the di-dee-do song to his wife and daughter, his trademark tune for when he relaxed. Nothing worried him at all. They were moving and making progress, the way he'd always operated. If he'd not been a mover, he'd still be working in a logging camp. Sometimes a guy just had to get going to get anything done. It was a gloomy day for sure near Portland, but not bad enough to stay grounded watching the rain fall while the overhead costs sucked up capital.

Like Homer's siren call that led ancient sailors to the rocks, each airport along the way promised clearer conditions. The promise of improved weather chummed the plane and pilot to the south.

Carla thought back to all she'd left behind: her dog, her horse, and her boyfriend. Phyllis looked forward to some warm weather. Up in the copilot's seat she helped Al with the navigation. In the back, Carla cuddled up under the extra blankets.

A thinking pilot leaves an escape route in case the weather drops. Ahead of the plane's bulbous nose, the only sketchy area lay between Medford and Red Bluff where the front passed earlier. This sparsely populated and rugged section of the route provided few airports, navigational aids, or communication links. If something happened there, they'd be in trouble, but Al calculated the odds and they tilted in his favor. The plane carried lots of fuel, the engine ran fine and the leg was short; they could cover most of it in an hour. For a short section of high mountains, The Viking would proceed.

Al calculated the odds and they tilted in his favor.

When air mixes and swirls, combining with moisture and differing temperatures, weather forms. When the brew occurs over high country, forecasts are often just crap shoots. Mountains are known to make their own weather. Al confidently planned on blasting through this mountainous route segment. Getting the winds aloft for 12,000 feet earlier showed his thoughtful strategy. As he approached the California border, he pulled the plane's nose up, climbing higher, knowing the elevations rose on the next leg. Dutchman Peak topped out at 7,400 feet so he'd at least have to clear that height.

Ahead, the clouds had built up into mighty columns into the skies. The Viking could weave her way through the shafts. Al knew that he must stay out of the imposing pillars. They contained deadly ice and turbulence. Flying past them was akin to walking through an old growth forest, altering the path to avoid colliding with a tree. Up front the smooth running Jacobs engine moved the plane swiftly along without skipping a beat.

Last Position Report

Oien last reported the aircraft's position near Medford, Oregon "in the clear," finally above the cloud tops at 11,500 feet. Being in the clear was like flying in a different world. Now no worries existed about striking a hill in rainy air. A few minutes previously he'd called the flight service specialist and reported The Viking five miles to the west of the airport at 9,500 feet in heavy rain. They'd now vault the abyss, the forbidding mountainous stretch between Medford and Redding.

Al stepped up to the challenge, continuing toward the Bay City via Red Bluff by climbing and using the height to his advantage. Right then over the radio he heard the final clincher on the correctness of his decision to continue south. The flight service specialist relayed a pilot report from a Beechcraft Bonanza over the Redding Airport saying the cloud tops were at 10,000 feet. Al knew then with certainty The Viking could get safe passage from Medford through to Redding. If a Beechcraft Bonanza flew over the tops at 10,000, The Viking would easily top those heights too.

The area where Alvin F. Oien last called to flight service is notorious for icing conditions. The famous California pilot Johnny Moore wrote about a frightening encounter with ice in a Cessna 206 while flying an emergency medevac into Medford from the south, the very area where The Viking began to encounter danger. In Moore's book *I Must Fly!*, he described the event which occurred on April 3, 1967, his 27th birthday:

"Suddenly we were all hanging from our belts in the severe turbulence with jolts so hard you could not see the instruments. Outside it turned dark with a greenish hue and I struggled to get my panel lights on. The 206 was quickly becoming a cake of ice, heavy stuff that was building at a breathtaking rate. We were virtually falling out of twelve thousand feet. I called center and told them as much as we slammed our way down through ten thousand feet. I had difficulty hanging onto the microphone. Lightning flashed and I exercised the propeller to loosen up the ice that flew off, crashing into the wings. The leading edges of the wings were protected by five inches of ice, so chunks of ice flying off the prop were not going to hurt them, besides, who gave a damn right now! I asked the controller to say our position and was told twenty-five miles south of Medford. Despite having every control pushed to maximum available power, we continued up and down, mostly down, and onward to the lower terrain that I knew lay ahead… Had we been many more miles south when this happened, no doubt we'd have hit the ground."

The Viking chugged along past the area where ice almost downed Moore's Cessna 206, a plane with similar performance to the Cessna 195. The predicted clearing period failed to coincide with The Viking's arrival, but the Sacramento Valley beckoned so close now and they'd flown so far. The little Cessna dodged and weaved around the gray clouds, staying in the clear. The fuel tanks carried plenty of fuel and they were sure to make Red Bluff where they could take a break and add more fuel.

A few hours after that, a gray and blue Cessna with a red rose on the tail would taxi up to the Butler Aviation ramp at the San Francisco International Airport. Al's family would then board the Delta jet to Texas and they'd all be on their way to better times. The Old Man never knew of his eldest son's plan to meet him at Butler Aviation. While The Viking flew on past Medford, climbing high, Delta Airlines First Officer Alvin F. Oien, Jr. waited for the plane's arrival at the San Francisco Airport. To pass time he watched planes and counted the minutes. At any moment he expected to see The Viking land, after which he could surprise and delight his father.

CHAPTER 5

Trouble Brewing

THAT AFTERNOON AT 1500, Al Jr. waited for the plane's arrival at the San Francisco International Airport, pacing around the ramp at Butler Aviation. He'd jump seated up from LA in the morning. Only Phyllis knew the The Boss's son would be there to meet The Viking as it landed. Junior looked forward to surprising the Old Man and Carla, whom he barely knew. Later, if the flight loads permitted, they'd all fly on Delta Airlines back to Dallas on the same flight. In any case, the Old Man's family had tickets, compliments of Junior through his employer.

In Dallas he'd host them all at his house with the swimming pool. In Texas the air stayed warm through March and a spring day in Texas was like a summer day in Portland. He looked forward to helping pull back the Old Man's finances from the precipice of disaster. The thieving manager from the Baker property put the hotel business in jeopardy to the tune of six figures. Over the years Junior put away a few bucks and studied business; he could certainly brainstorm with his dad to find a solution to the mess.

After years of being like oil and water, the two respected each other. Al Jr., whether he admitted it or not, inherited some of his stubborn drive from the Old Man. As a young man, Junior refused to have anything to do with him, but an Air Force buddy always urged him to reconcile with his father, "no matter what." Junior took the advice to heart and faced the pain of making peace. The two men started talking in Portland years before and found much in common. They were both pilots and they both studied and enjoyed the challenges of business. Junior found out too there were two sides to every story and began to understand his father's marital problems much better.

The anticipation of a big surprise and happy reunion dissipated as the afternoon wore on without word from the N9388A. No real need to worry when dad flew, Al Jr. thought. San Francisco weather stayed clear enough while the hours passed. Junior figured, as he told a reporter later on, his dad had "set down somewhere along the line and was drinking a cup of coffee," waiting for the weather to improve.

Forty years later, the son recalled the afternoon when the plane did not show up. At Butler Aviation he walked the ramp, watching planes. A pilot pre-flighted a personal jet in front of the hangar, planning an evening trip to Las Vegas for the sole purpose of the jet owner eating dinner there, an amazing idea to the young pilot. "Imagine, just flying a jet to Vegas for dinner," he thought at the time. Junior killed more time, waiting to hear from Oakland air traffic controllers who'd phoned him when The Viking failed to arrive on time. The lack of word failed to bother him. He went to a hotel and rested for the night, still confident all was well with his father's flight. Surely by morning he'd hear some word from the family about their failure to arrive. Most likely, the old Viking suffered a mechanical malady or the weather had interfered.

Junior figured his dad had set down somewhere along the line waiting for the weather to improve.

INREQ

Fortunately the Old Man filed the flight plan when he left Beaverton that morning. The filing ensured if something went awry, help could be dispatched right away. When The Viking failed to make San Francisco at 1500, a series of actions occurred after a one-hour grace period. At 1600, The Oakland Flight Service Station issued an information request over the FAA teletype system. This notice, called an INREQ, called for any FAA radio station records of contact with the overdue aircraft. As a matter of routine, the FAA forwarded INREQs to the duty officer of the 41st Rescue Squadron at nearby Hamilton Air Force Base. Keeping the Air Force apprised of looming search possibilities allowed quick mobilization of search planes and parties.

The 41st carried search and rescue coordination responsibility for a large portion of the western United States. Known as the Western

Aerospace Rescue and Recovery Center, or WARRC, the base operations worked extensively with state agencies, volunteer groups, local law enforcement, and the Civil Air Patrol (CAP) to locate missing aircraft. Because of the vast and rugged landscapes of the west, SAR coordination proved to be a huge challenge, as it is now. In spite of bountiful resources, people sometimes just disappeared forever. In 1967, over 31 aircraft containing 59 people lay somewhere over the nooks and crannies of the WARRC's area of responsibility: Washington, Oregon, California, Idaho, Montana, Nevada, and Arizona. There were no emergency radios to pinpoint crash locations. In 1967, aircraft rescue beacons existed only as a technology of the future for civilian pilots. To be rescued, a downed pilot had to be spotted, heard, or trek his own way out of the wilderness. If none of those things occurred the pilot just remained missing.

Then, as now, pilots had to stick to their flight plans for search and rescue resources to be dispersed properly if their planes became overdue. The flyer who deviates from a pre-planned route due to weather or some other factor should relay the change to the nearest ground station as soon as possible. If a pilot fails to report a change in flight plan, the search may focus on the wrong area. At least Al Oien stuck to his flight plan, staying close to the airways between Portland and Red Bluff. The plane meandered off the airway by a few miles while dodging cloud buildups, but lay close enough to the airway to be spotted when the search planes took off.

Resources are diluted and planes can remain lost forever if a search area is too large. The value of a flight plan is that it provides a clearer location for search and rescue units to start looking. Obviously, when injured people await medical treatment, or impending weather threatens, a focused search provides the best chance of being found alive. The odds of survival for the downed aircraft survivor decrease as each hour ticks past.

From the start, searchers examine the pilot's habits. What is their flying history? Are they risk takers? What is their track record? Have they been lost before? What training and experience do they possess? What about aircraft capabilities and equipment? These are the sort of questions that were asked about Al Oien and his plane.

Oien's flying record carried a minor history of scrapes with the aviation authorities. First he lived through the Chehalis crash when he damaged his back and wrecked the new Luscombe. In the mid-fifties he landed at Pearson Airpark in Vancouver in the Fairchild 24 without landing lights. That night, he just wanted his wife to see the lights of Portland, so he took her for a scenic flight. Car headlights worked just fine to illuminate the runway. Unfortunately a Washington State Patrol officer, who happened to be a pilot too, saw the Old Man land illegally. The officer arrested the pilot. Ultimately, the courts convicted Al and he lost his flying privileges for one month in Washington State, a precedent-setting decision because states rarely regulated pilot licenses. Also, back in the 1950s on a flight in the Fairchild 24 from the Midwest through to Oregon, Alvin Oien became the subject of another search by the CAP, though he did nothing wrong. His flight route through La Grande changed due to weather, but the change failed to be forwarded with his flight plan and searchers launched in Montana while Al slept in his plane at La Grande airport, waiting for better conditions. In the end, Civil Aeronautics Association, or CAA, determined he had actually notified a CAA ground station of his changed route through the La Grande Airport Flight Service. The confusion had stemmed from poor communication between CAA stations. Al had a "history" in his flying for sure, but the bottom line bore out the fact that a very capable pilot flying in a very capable plane was missing.

ALNOT

While the INREQ buzzed through the FAA communications system, minutes passed with no news of the plane. A long hour later, Oakland Flight Service issued an Alert Notice (ALNOT). FAA tele-type machines chattered with this higher priority signal, which gave immediate notification to the 41st Rescue Squadron Duty Officer that an aircraft had likely gone down somewhere in the 41st's area of jurisdiction. Trouble brewed out to the northeast, it seemed, and the Air Force may as well get ready for a mission.

The fact no station reported recent contact with The Viking raised the level of concern. The ALNOT found the last known location of The Viking. Close to 1100, when the aircraft passed just west of Medford, the Old Man talked to the Medford Flight Service, reporting the plane above the cloud tops. No one had heard anything from the Cessna since then, and it never arrived at Redding, Red Bluff, or San Francisco.

If the pilot of N9388A elected to fly back north surely he'd have called the Medford Flight Service and reported the route cancellation, unless he diverted to the west or east. Based on the ALNOT, officials believed the plane was down somewhere between San Francisco and Medford, hopefully parked at an airport. The rescue system slowly mobilized on Saturday. Search efforts started from the Medford area southward. At the beginning people checked each airport along the route. Others compiled information about the pilot's history and habits along with the aircraft's capabilities. Mechanics and pilots of the 41st Rescue Squadron inspected their Grumman HU-16s. It looked as though they'd be flying into the dark night unless the pilot of N9388A checked in soon.

At 1800, as the daylight faded, two twin-engine Grumman HU-16 amphibians of the 41st roared their way northeast from Hamilton Airbase to check the country along the airways between San Francisco, Redding, and Medford. The search for the Oien family officially began as the echo of the powerful engines of the silver and yellow amphibians bounced through the lonely mountain canyons. Early on, one Grumman returned to the base with engine trouble.

Into Sunday morning the big plane flew over the peaks along Victor 23, the imaginary line that connected Redding and Fort Jones and Medford for aviators. Below the wings lay nothing but the dark and wildness of a mountain getting buried by an immense snowstorm. Since nothing but blackness blanketed the mountains by the time the Air Force planes launched, the HU-16 checked for fires and distress signals. With their radios they listened for distress calls. The aircraft carried two pilots, a navigator, flight engineer, and two Air Force Para-rescue specialists: men capable of parachuting to a crash site, rendering aid, and directing ground rescue.

The 41st Squadron Duty Officer called Al Jr. at his hotel, informing him of the decision to launch. Getting on toward dark, there'd been no word from the pilot or his plane. Alvin Jr. knew the search and rescue drill, how the first hours of a search were critical. He could write a book on search and rescue. His thousands of hours flying the SA-16s for the Air Force at Cape Canaveral, then later for the Air Force Reserve up in Portland, had burned the proper procedures into his brain. He knew exactly what the Air Force pilots were doing that night; the HU-16 was virtually the same as an SA-16.

As the Grummans covered the possible track of the missing plane, the ground teams swept through the airports of Northern California and Oregon, checking registration numbers of tied down Cessna 195s, searching specifically for a gray and blue one with N9388A painted on the side. AM radio stations broadcast The Viking's description and route, urging listeners who heard or saw such an airplane to call the local sheriff. By day two, even a team climbed the sides of Mt. Shasta, following up on a report of a downed aircraft.

The ocean moved in from the west and unleashed a furious blast of rain, snow, and wind over California, just at the time of The Viking's disappearance. This weather was exactly what Al had conscientiously tried to avoid by leaving Saturday morning. Now the family was in trouble. Over the next four days, snow fell without respite over the high country of the west. Where the clouds did not dump snow, the rain, wind or ice destroyed the routine of daily life. In the Sierra Nevada, many feet of snowfall stranded thousands of people. There would be no worse time to disappear over the Northern California mountains in an aircraft than during the week of March 11, 1967.

There would be no worse time to disappear over the Northern California mountains in an aircraft than the week of March 11, 1967.

By Sunday, facts established The Viking had put down in the wilds somewhere between Portland and San Francisco. Experts believed the plane made it over the Oregon border into California. By 3 p.m. the previous day, her fuel tanks burned dry. Somewhere out there, three people needed immediate help or were dead. No one could do anything except wait for the storm's passage, and the weather forecast maps showed nothing but poor conditions for the immediate future.

Air crash survivors in the wilderness don't live long, even in good conditions. Ideally, marooned fliers are found in what some call the "golden window," the first 24 hours of the accident. Facing injuries, privation, exposure, and shock, human lives usually last only for a few days in the wild. Three days is the average time an uninjured survivor stays alive after a crash on mountainous terrain during winter. To keep breathing beyond that is a mixture of luck, will, preparation, condition, and supplies. Some people beat the odds, but most don't.

Initial reports showed the 41st Rescue Squadron flew their planes along the airways from San Francisco to Redding to Medford for the rest of the weekend. The search pilots detected no sign or radio signals from Oien's aircraft. As the weekend rolled past, the duty to find the Viking shifted to the Civil Air Patrol and the 41st Air Rescue Squadron returned to standby status back at Hamilton Air Force Base. The Hamilton planes flew one more sortie during this search, investigating notice of a crash site reported near Mt. Shasta.

Since few CAP outfits were based in sparsely populated Northern California, most of the pilots, dispatchers, and ground crews were imported from units elsewhere in California. Historically, when the weather dropped in the mountains, nobody flew much in civilian planes. The high terrain limited the radio navigational aids' range and the barren country limited safe havens for aircraft during bad weather. Pilots usually tied down and waited for good conditions before tackling flight through the treacherous country. Valuable time passed as the small search planes staged for the mission; this hampered the search from the start.

The golden window of life for air crash survivors had closed already as the CAP ramped up for a full-fledged air search, focusing specifically south of Medford, the last known position (LKP) reported by the Old Man. Whatever happened, most likely the treacherous maze of the California-Oregon border-lands held the secret of the missing plane. Using a budding technology of estimating icing via weather radar, it was confirmed that a Cessna 195 could not outclimb the icy clouds over the border country at mid-day on March 11. CAP planning officials methodically laid out search grids on a map of the expanse referred to as the area of probability. The Cessna 195 had to lie somewhere in the CAP map grids. They just needed to launch their fleet and start looking once the weather cleared.

If there'd been a search planning tool that took tenacity into account, they'd have quadrupled the size of their search area.

Unfortunately for all involved, when making the assessment of the pilot's qualifications, experience, habits, and plane capabilities, the search planners missed an important aspect of the pilot's habits and personality: The Old Man did not understand the word impossible. The concept did not seem to exist in his vocabulary. If it was clearing cloud tops in his Cessna or cutting himself from a body cast so he could fly to Cuba, the Old Man pushed limits. If there'd been a search planning tool that took tenacity into account, they'd have quadrupled the size of their search area.

CHAPTER 6

Crashed!

Searchers could look at all the airports in the world and never find The Viking safely tied down at any ramp with The Boss sipping coffee and waiting for the weather to clear. The gray plane with the red rose on the tail, which left Portland in style just a few hours before, now lay crumpled and smashed at the 4,300 foot level of Shoemaker Bally Mountain in the Trinity Mountains, beneath the heavy falling snow.

Her broken nose now pointed due east. Her wings with their rounded tips lay against the steep slope, left wingtip pointing downhill. Her back cracked where the strutless wing squished down into the fuselage. Icy mountain wind filled the broken plane. In the end, The Viking dropped from the sky like an overloaded elevator, and the Old Man weaved her down through the towering peaks like someone who really knew how to fly. With the flaps for max lift and the throttle for best power he finally brought his airplane to a stop without killing anyone, a tremendous feat of piloting.

He did what a pilot should do when shit hits the fan—he flew that airplane until it could not fly another inch. In his life there'd never been any flinching, and on this last flight, as proven by his smashed face, there'd been no ducking or hiding either. Maybe he radioed for help as the plane dropped from the sky, but the mountain peaks probably blocked the signals. No one ever reported hearing distress calls. It is speculation to say exactly what occurred in the cockpit when The Viking hit the mountain, but the post-crash events were well documented.

As it turned out, if only Al's plane had flown a little bit higher and a quarter-mile further to the west, the Cessna could have escaped unscathed over the Northern Sacramento Valley, allowing the Oiens to make the ETA into San Francisco, alive and well. Luckily for them, Cessna engineers did not scrimp on aluminum when making

Cessna 195. Because of the stout monocoque design, the strong shell of the cockpit stayed intact. Cushioned by the snow-covered and springy layer of brush, The Viking slammed the mountain hard enough that she'd not ever fly again, but all her passengers lived. First the wing hooked a tree and spun the craft around, transforming the forward motion into a powerful sideways spin. Then the whole plane hit the mountain slope in a spray of snow and the entire works shuddered to a stop before sliding backwards downhill, finally grinding to a halt on an impossibly steep snowy slope.

When the plane hit the ground, the forward windshield broke out on the right side, sending jagged pieces of Plexiglas through the air to cut and puncture. The empty windshield hole acted like a scoop. Snow, brush, and freezing air poured in while the strong single-piece wing twisted and cracked, leaving a gap in the back of the fuselage. The exit door emblazoned THE VIKING jammed closed from the weight of the wing, leaving the only escape through the smashed windshield's gaping hole. Delicate instruments and radios shook to bits, and gravity squeezed Carla, Al and Phyllis against their lap belts. The cool mountain air made the warm Jacobs engine cylinders tick as they lost their heat. The cockpit stayed silent except for the howling winds and blowing snow.

Only time could show if fortune had really smiled when the three survived the crash. Already the hands of time ticked down on their chances of living. The wet clouds that had piled ice on the plane's wings unleashed their pent-up force on the countryside, covering it with wind, rain, and snow. An air search could never launch in those conditions, and as each minute passed, the snow piled deeper and higher while the three unfortunate people in the wrecked plane got a bit colder and weaker and began their slide into eternity.

Just across the way from the crash site sat the hamlet of French Gulch with its handful of people. Weaverville, the Trinity County seat, lay less than twenty-five miles distant. Highway 299 ran west to east, only eight miles away, connecting Redding with Weaverville. During good weather, private planes crossed above the Trinities, taking a shortcut toward the Sacramento Valley. Above Shoemaker Bally Peak, up to sixty airliners per day flew between the Red Bluff and the Fort Jones VORs, navigating by the electronic signals defining the air highways. These aircraft cruised way too high for anyone aboard to see the tiny speck of a blue and gray plane against the vast mountain country spread out below, although they flew only a few miles above the crash. With no means to signal, the Oien family might as well have been hearing or seeing interstellar spaceships traveling between the stars, as the airliners droned above.

CHAPTER 7

The Diary

IT IS POSSIBLE THAT everyone in Oien's plane blacked out after the final impact with the mountain. Both parents suffered some sort of head injury. Phyllis Oien woke up delirious after the accident and Alvin suffered facial injuries that, according to letters, "gushed blood." Before she regained her faculties, the icy mountain air froze Phyllis's bare feet. At least that is what the diary recorded, the one she began on the back cover of the 1964 Airman's Guide, the book pilots use to keep abreast of changes in aviation related regulations.

Perhaps the diary began as an afterthought, but most likely, the reason for the diary was that Phyllis just liked to keep track of details. At home she'd carefully listed the contents of the suitcases carried on the trip; in her records she'd carefully documented lists of her possessions. There'd been the time when she even documented the possession of a broken mug handle. A list maker all her life, she decided to record their adventures in the crash, something to keep busy with before the air cavalry swooped down to their rescue.

On the Airman's Guide margins she'd write the daily details of the crash and of their struggle for survival. There was little else to write on. The 1961 bestseller *Fate Is the Hunter* by Ernest K. Gann, and a serialized magazine story about Amelia Earhart were the only other reading material on board the plane. Both were poor subject matter for people marooned in an air crash in the wilds. No pilot of The Viking would need the 1964 Airman's Guide anytime soon; it could not go anywhere and would not fly again.

The three survivors' reasonable mindset from the beginning seemed to be based upon the security of having filed a flight plan. Finding them should be a no-brainer. Surely with the aircraft on a flight plan, its path along an airway, and military bases around, they'd all be rescued soon. Going down in such circumstances was nearly akin to pulling off to the side of a busy highway to deal with a problem. Mistakes were made but Al had done the main thing well: flying the plane down to a successful forced landing without killing anyone. Now they just needed to be patient. Once the weather lifted, surely there'd be a search and rescue.

Just one hour previous to the crash, Al Sr. transmitted their plans to proceed southbound when he reported their climb to VFR conditions above the clouds to the Medford flight service specialist. To anyone with a sensible thought in their heads, the whereabouts of the crashed plane should be fairly obvious and easy to figure out. If it failed to arrive at Red Bluff, the location should be somewhere between Red Bluff and Medford, in close proximity to the airway. Finding it ought to be a fairly straightforward affair with all the airports and resources available in the area. Military airbases lay north and south of them. Military bases meant big, professional search planes and helicopters and good equipment—para-rescuers who could jump down with their medical bags, patch everyone up, and get them ready to be pulled out of the mountains by helicopters.

Practical, organized, take-charge Phyllis took care of their injuries as much as she could, then documented their situation in the same way she handily inventoried everything. Besides, writing would fill time until the inevitable rescue. With a red pen she wrote on the spaces along the margins in the guide book:

"1967 March 11 1215

Saturday March 11 on leg Jones omni to Redbluff

Plane on left side in snowbank. Fuselage broken. Door ajar. Windshield right side windows broken.

Injuries—Al, cut on chin, 3 cuts on forehead, right arm broken. Pain in vertebra, crushed ribs on both sides.

Phyllis, delirious one day. Broken left arm. Sore right hand, bruised left leg. Cut and/or broken left ankle. Both feet frosted first night. Chin and nose bruised.

Carla, hurt back above left kidney, Sore right ankle, cut on left knee.

On hand, M and Ms, sour orange drops, mountain of snow. Rearranged interior."

Competence and leadership came naturally to Phyllis Oien. She graduated from Northwestern University with a Bachelor of Science degree when few women even attended college. When her husband failed to live up their marriage vows, she sent him packing and made a life for herself and young daughter. Whether it was running a scrap metal drive for the war effort or joining the Army Nursing Corps because the country needed nurses, she always put herself forward to complete the task at hand. The ultimate proof of her tenacity was this: She made history as the first person in memory that could even begin to control the Old Man, something plenty of others tried to do but failed.

Phyllis forged her character through pain. Life challenged her after the disintegration of her first marriage due to her husband's love for drugs. She prevailed through the trials of single motherhood while raising her daughter and working full time at her professional nursing job. Although grievously hurt in a remote crash with broken bones, cuts, and bruises, she took charge of her dazed and fractured family, patching up their injuries with whatever she improvised from the limited contents of the wrecked plane's luggage and tools.

Enduring the terrible first night of the crash, sometime during the first twenty-four hours the three rallied enough to rearrange the interior of the plane so they could be more comfortable. They "tore it apart," giving them a space to huddle in not much bigger than a coffin. Resourcefully, they secured their shelter from the elements. Maps and pieces of the plush Cessna *Businessliner* interior plugged and stuffed the cracks and gaps of the broken cockpit, sealing it from the nonstop cold mountain wind. The aircraft rested precariously against the slope in such a manner the left wing pointed downhill, so the left side of the cabin became the floor, while the original floor became the right side wall.

They opened and checked the contents of the suitcases, gathering clothes and blankets. Fortunately they carried the two blankets along with some thick homemade sweaters. To cover the broken-out right side windshield and block the gaping hole from the freezing blizzard, they used the heavy canvas engine tarp. This canvas normally protected the cylinders from the elements if the plane were tied down overnight. The windshield hole was now the only access point in or out of the plane. The broken wing, which had cut into the fuselage, blocked the right side entry door. Each exit and entry through the windshield hole required someone to move the heavy tarp. Painfully injured, movement in the cramped space caused severe discomfort.

The diary in Phyllis's hand noted:

"waterworks, tastes exquisite, yielded about 1 gallon."

"Three days of heavy snow. 1 day mixed, 1 nite rain."

In spite of his injuries, Al rigged Carla's bathing cap into a water collection device. His ingenious invention became known as the "waterworks," a jerry-rigged contraption used to trap drinking water from melting snow. As the waterworks poured out the life-giving fluid, the family felt a sense of accomplishment.

Later they wrote about the relief of getting a drink—one of the only bright spots of the entire ordeal. Adequate water is even more important than food for short-term survival. Using their brains to solve a problem, helped them feel better.

Intelligently, they took care of their basic needs just like with the waterworks. First they arranged a good shelter, cared for wounds, took stock of their supplies, and kept a positive outlook, hoping for a rescue. With foresight, they rationed their limited food supplies, nothing more than lemon drops and M&M's, six per day per person.

In an ideal situation, downed plane survivors mark their aircraft so it contrasts against the surrounding terrain, allowing rescuers to easily spot it. Messages stamped in snow and brush formed to spell words, signal fires from aircraft tires and oil that release thick smoke: all have been used to mark wreckage or signal rescue parties. As for The Viking, there'd been no point in marking anything during these first days: The endless cloud cover ensured no one could possibly see an airplane against the white background of the slope. Nor could the downed people signal upwards through the overcast. The wind and blowing snow made standing outside like being in the path of a freezing blowtorch. With twenty feet or less for visibility, going out-side meant getting lost, freezing, and dying. While the storm raged, hunkering down and enduring the slow passage of time made sense.

With a mountain of snow for water, some hard candies, M&M's, and a few small jars of jelly existed to nourish their bodies; they just needed to hang on for a few days. Surely they'd be found once the air cleared.

They knew too that Al Jr., the former Air Force search and rescue pilot, would never stop looking for them. The three were tough and smart folk who'd thrived in their previous lives in Portland. All shared the same love of outdoor pursuits: camping, gardening, fishing, flying, and photography. Their group intellect carried enough knowledge, energy, and wisdom to get through just about any ordeal. The factors against them were time, injuries, little food, poor shelter, and worst of all, the searchers were focusing on and looking in the wrong places.

Worst of all, the searchers were focusing on and looking in the wrong places.

Weather for the upcoming week promised nothing but trouble. In 1967, without Emergency Locator Transmitters (ELT) to guide searchers to a lost craft, search fliers had to see every square inch of the country. Air searches required clear air, an impossible requirement with the thick clouds shrouding the west. The Civil Air Patrol (CAP) had made the plan and positioned aircraft and people in strategic points, preparing for a clear break in weather. Once the cloud cover lifted above the mountains, the planes would swarm like bees over the area of probability.

By late Sunday, clouds still grounded CAP planes in places like Chico and Yreka. Widespread low ceilings, reported to be 600 to 1,200 foot over the ground, kept these planes from viewing the higher country along The Viking's route. Frustrated and knowing lives hung in the balance of the rapidly passing time, CAP pilots stalked the airfield ramps, peering into the gloomy, wet skies. There'd already been a failed search the previous December for pilot Art Ward and his *CallAir* after the Washington pilot fueled his low-winged ag-plane in Dunsmuir and flew off to oblivion. Ward and his aircraft were still out there somewhere. Up north near Medford in slightly better flying conditions, two civilian search planes glimpsed the lower parts of the Siskiyou Mountains. The CAP spokesman claimed if weather cleared, the area could be completely checked by Tuesday. After that, the CAP planes would move further south along the area of probability.

The CAP protocol considered an area searched once an aircraft and observers scoured a sector a minimum of three times. A pass or two could not suffice for one hundred percent coverage.

Nevertheless, the efforts became a mostly hit and miss affair due to rapidly changing flight visibility. Sometimes a mountain allowed complete scanning, other times it clouded up and needed to be searched once more.

The Civil Air Patrol is a civilian branch of the United States Air Force formed in 1941, in the face of an impending war. Conceived to supplement the military forces by using the thousands of civilian pilots and aircraft in the U.S., the patrol maintained an admirable history of service. During the war, CAP pilots flew anti-submarine patrol missions, search and rescues, and border patrols, among other duties. One CAP plane is credited with sinking a German submarine. After World War II, the organization assumed domestic search and rescue duties and other congressionally mandated missions.

The Civil Air Patrol is composed of state wings, with local units broken down into squadrons. The units are composed of pilots and ground personnel who fly private and government aircraft. The CAP organization is credited with saving on the average over 100 lives per year. Because of the dangerous nature of search flying, hundreds of the organization's pilots have lost their lives since the inception of the patrol.

Al, Jr.'s Search Begins

While the search and rescue system mobilized, Al Oien, Jr. knew he had to do something more than wait on the sidelines. Like his father, Al Jr. grew up into a man of action who did things his way. As the eldest son he felt a duty to find the family, plus he felt a bit responsible for the accident: He'd arranged the tickets, the trip to Dallas, and had the ideas that could pull the Old Man from the brink of bankruptcy. After a long, sleepless weekend waiting for word of his father's aircraft, the younger Oien took a leave of absence from Delta Airlines and began planning his own mission. He'd get his old friends to help him.

As a former military pilot who had specialized in search and rescue, he associated with men who were mission-oriented; pilots who focused on and completed difficult tasks. His background as a professional search and rescue pilot for the Air Force caused him to avoid the CAP. He did not like what he saw with the civilian pilots and their puddle jumping planes. To him, the amateur pilots expressed good intentions but worked under misconceptions.

For one thing, Al Jr. noticed the CAP claimed the one hundred percent coverage of a map sector. In his mind and experience, such a claim could not be possible. If the authorities made their final decision to stop searching an area based on such faulty methodology, the possibility existed they'd miss a plane. He knew aircraft, people, and sought-for objects turned up often enough in previously searched areas. Sectors that had been searched "thoroughly" only meant no one had seen what they were looking for. No areas should or could ever be called one hundred percent covered until the missing object or people were found. Right now his dad's family lay missing somewhere out in the wilderness, and people systematically checked off small squares on a map saying they were not there!

Sectors that had been searched "thoroughly" only meant no one had seen what they were looking for.

Besides the headaches of the search and the heartache of the missing family, the problem of the Clifford Hotel and the Old Man's other properties required immediate action, and there was little to go on in the way of records. The Old Man kept his financial papers in his fedora and the income numbers in his head, making it a damned complicated business to deal with when neither the man nor hat could be found. He'd never committed to a will, though the year before he'd met with his attorneys and begun estate paperwork. Phyllis, following her practical ways, insisted on estate planning, but Senior laughed it off. Like many people, he knew he'd never die soon. It wasn't just the Clifford Hotel's day-to-day needs of payroll, maintenance, and guests that posed issues; the impending legal action on the white elephant of a hotel in Baker, Oregon, also reared its ugly head. The thieving manager left the business in a precarious state. If the Oien boys did not take action, their father could lose everything while he waited for rescue.

Al Jr. recruited his younger brothers, Ron and Chuck in Portland, and they dropped everything to attend to their father's complicated business affairs. Like their older brother, the boys had suffered under the wrath of their Norseman dad. Later in life, when they watched the movie about Patrick Conroy's *The Great Santini*, they laughed, Santini being a great wimp compared to their experiences with their father. Ron and Chuck would keep the Clifford Hotel open; too many people depended upon the building for it to close the doors. There were guests, business owners, and the live-in tenants who relied on the building. Now the man who'd insulted them so much sat on a mountain somewhere in Northern California and desperately needed their help. The boys held the cards, and though they could walk away like the Old Man did to their mother, they now had a big chance to prove to him they were made of the same tough blood. The Old Man had cultivated loyalty with his employees, and a benevolent manager by the name of Dick Purdy gave the boys a crash course in hotel management. When the Old Man walked out of the woods, they'd show him a thing or two about how to run a hotel by having kept all the doors open.

Dick Purdy had been the go-between friend of the Old Man, keeping him apprised of the status of his sons. A founder of the Purdy Brush Company in the Portland area, he had seriously hurt his back in an accident. The Old Man had suggested Purdy take over the night manager slot at the hotel since he could not sleep well and had to stand up for long periods of time.

The Old Man's belligerent ways with his sons and family never carried over to his employees. Hotel people loved working for Al F. Oien, Sr., and they felt his absence keenly at the Clifford when he disappeared. Later on, the employees all signed cards with heart-wrenching notes of sympathy.

Al Jr. knew exactly what to do to find The Viking. He called up his old Air Force search and rescue unit, the 304th Aerospace Rescue and Recovery Squadron, where he served as a pilot when he left active duty—and served with honor. Once he even located the crash of a U.S. Navy P2V Neptune on Mt. St. Helens in Washington, the final flight of a Navy Admiral's son. The night of the patrol plane crash he'd been the duty officer who triangulated the stricken aircraft's estimated position to within a few miles. Searchers pursued his suggestions and remains of the Neptune were recovered.

Junior came up with a plan and pitched it to the Oregon Air Guard. The Squadron Commander saw things Al's way. If the pilots required training, it may as well be on a real mission. He assigned training flights using the missing plane instead of using simulated search objects, and allowed Al to fly along.

Time was now the critical factor. The weather maps reflected the horrible pattern that so far kept the search planes down, but by using professional search pilots and heavy planes instead of the civilian pilots and planes of the CAP, maybe Al's family would be found. The military aircraft were well suited for search flying, being capable of long duration flight. They were the Grumman SA-16 Albatross, just like the Hamilton-based planes. Al Jr. rode along as an observer in the Portland-based SA-16s while the CAP flew the smaller planes from places like Medford, Yreka, and Chico.

Years later, Al's recollection of the search bothered him. His image of the time aloft in the SA-16s was of tough flying in lousy visibility. With the storms piling in wave after wave from the west, he estimated less than one percent of the terrain was viewed accurately. He shook his head and could not understand how the CAP could state with such certainty that an area had been completely searched. Just like with the small CAP planes, as the clouds cleared, with the ungainly SA-16s positioned for another low pass over their map quadrants, more ugly clouds moved in to obscure the observers' view. Sometimes they'd be forced to circle over the area until the overcast blew past, if it ever did. The 304th ARRS continued the search flights until the end of the official search, loading up the Grummans each day with observers and flying to the northern California mountains from Portland.

As news of the missing Oiens spread through the West Coast newspapers, in Portland, Carla's boyfriend Rick discovered his girl never made it to Texas as he thought. Instead, her father's plane crashed somewhere near the rugged Oregon border. Using his last twenty dollars for bus fare, the high school senior rode to Medford where he took up a cot in a hangar at the Medford Airport, spending sleepless nights and heart-crushing days.

Too young and untrained to fly as a searcher, he anxiously paced the airport ramp, meeting each search plane as the pilots taxied in from a flight. The negative shake of each aviator's head told him all he needed to know: that somewhere the girl he loved was in a crashed plane lost in the rugged mountains. If only he could fly, he'd find her. He and Carla had the secret marriage plan they'd announce on her 16th birthday, coming up in just six weeks, on the last day of April and his hope for the future burned bright.

DAY 3

Monday, March 13, 1967
   Diary:

   "Monday Carla tried to walk out. Returned. Snow, fog.
   No visibility. Snow 3 feet deep. Feet frozen. Shoes lost."

The trio endured another freezing day while over 36 inches of snow buried their airplane. The deep snow made it impossible to keep the wings clear. No one could see them through the cloud layers anyway. With the fuselage sealed as well as possible from the weather and the waterworks supplying their fluids a gallon at a time, they did not have much to do but wait, rest, heal, and plan. To ward off the cold, each wore up to seven layers of clothes against the temperature of the mountain air. Since the accident there'd been nothing but heavy snow and roaring winds or fog at the crash site. The air never warmed above freezing. Lying in the cramped fuselage resembled living in a crumpled freezer. As they shivered they burned precious calories, watching their breath frost up the interior, coating everything. Active adults need thousands of calories per day just to maintain their body functions, even while at rest. Cold people need more calories. In icy climates, fat is an efficient fuel. Arctic explorers are known to consume butter by the stick with no weight gain or cholesterol increase. The low temperatures make the body burn the fat's energy the way a hot stove burns fuel.

   The Oien's mountain pantry carried no fats, only the handful of candy, the few jars of jam, and the piles of snow. Smartly, they'd limited their food rations from the beginning. This meager amount of energy guaranteed nothing but extended privation. As time passed and the food dwindled, their bodies would eat themselves in order to survive.

   Meanwhile they had to have pondered their situation, running it over and over in their minds. Once again, the aircraft crashed near civilization, along a well-traveled airway. Airbases with search planes were nearby and the Cessna's route was detailed on an official flight plan filed with the Flight Service Station. The flight plan served the very purpose of locating an aircraft should it not make it to the filed destination. Al had just called their position into the Medford station an hour before the accident, so their whereabouts could be narrowed down to a simple stretch of the course line off a bit more than one hundred miles. Surely a large search effort was underway. This was 1967 and people just did not disappear in plane crashes in the United States. Weren't survivors always found?

Air crash survival experts recommend survivors stay with the plane in the event of a crash. A downed aircraft is a much larger object than a small human being desperately walking cross-country. The airplane provides a much better chance of being spotted.

In Monday's storm only a hardy and desperate person would dare to venture outside the fuselage's shelter, and that was Carla, the least injured of the three. Visibility in the blowing snow was down to zero feet and the wind chill hovered at arctic temperatures. Movement by foot across the steep mountain slope was almost impossible under the best of conditions, making it doubly difficult during the winter with the snow and poor visibility. Underneath the white layer sat a latticework trap of springy, brushy chaparral and scrub bushes ready to trip the unwary. The vegetation that cushioned the plane to a soft landing, now proved to be a formidable obstacle to cross-country travel. Leaving the fuselage and moving along the slope was like walking across a flimsy snow-covered tightrope of vines and scrub. If the traveler's foot failed to find secure footing, it plunged through, sometimes leaving the person flailing in waist deep or even armpit deep snow.

Complicating matters was the steepness of the mountainside. It seemed downhill was the only direction of movement possible in the deep snow. The mountain's uphill slope presented an impossible challenge without mountaineering gear. As it rose above the 4,300-foot level, it became almost vertical. Covered with snow until it could hold no more, the pitch offered the further danger of avalanches.

Her stomach ached from hunger. Like a live monster lurking underneath the snow, the brush vines grabbed at Carla's feet as she floundered away from the safety of the cockpit. The brush won by claiming the shoes off her feet during this first expedition. After stumbling around the steep mountainside, burning precious body energy, she found her way back to the relative safety of the plane and her parents. She would do one more hike outside the safety of the plane, but after that she'd spend most of her time waiting inside the plane, the only possible thing she could do with frozen feet. Later she outlined the ordeal in a letter to her boyfriend saying "to hell with it, I'd rather die with mom and dad."

Frying Pan Country

The brushy and impossibly steep land of the Trinity Mountains is known to local hunters as "Frying Pan country." In those parts it is said that if you shoot a deer away from a road you'd best be carrying your frying pan and be able eat your animal on the spot, because there is no way you will be packing it out through the thick Brillo pad of brush.

The bushes of frying pan country that cushioned the plane were the same brushy vines that made cross-country foot transit difficult. The Viking had come down smack in the middle of frying pan country during the coldest, wettest, snowiest March in years. On top of the record-setting precipitation, there was much more wetness to come. Over the next days, some places in the Sierra Nevada reported up to fifteen feet of powder snow. The sides of Shoemaker Bally took their own load of heavy snow while the San Francisco Bay area suffered under rainfall measurable by inches. Flooding, landslides, and icy roads took lives throughout the area during the big storm's reign.

The bushes of frying pan country that cushioned the plane were the same brushy vines that made cross-country foot transit difficult.

The monstrous low-pressure system that caused all the trouble still laid coiled like a snake off the Washington shore. It slowly lost its power by expending itself in the series of violent impulses that peeled off each day or so, hammering the coast. Though weakening, the system remained in place. Precious time ticked away.

On Monday, as darkness replaced the daylight, the heavy snowfall continued. Only one thing remained certain to the small family: They'd be spending another night in their cockpit deep freeze. No doubt they laid awake, their breathing expelling cloudy vapors on each other as each thought of home, warmth, food, and better times to come, as well as in the past. The search pilots, those few who could get out in the weather, returned to their bases, frustrated knowing people needed their help.

DAY 4

Tuesday, March 14, 1967
 Diary:

"Tues. Heavy snow, boom heard 1115 each day"

Planes passed above them on the aerial highways throughout the days and nights. For the Oiens there was really no choice but to wait.

Absolutely nothing could be seen through the clouds. The entire Trinity Mountain Range lay covered beneath the thick, gray masses. Even if Al Oien, Sr. lit the whole mountain on fire with the leftover aviation fuel, the blaze would remain unseen through the heavy weather.

The mysterious explosive noises were not related to any search activity for The Viking. Investigators traced the puzzling sounds to departures of the supersonic SR 71 Blackbirds which flew spy missions from nearby Beale Air Force Base just to the south of Shoemaker Bally. The SR 71s carried cameras which easily could have photographed the rivets on The Viking from near space, but the capabilities of the famous Blackbirds were used for only spying.

By late Tuesday, the Civil Air Patrol and Oregon Pilots Association announced the Medford sector search completed with one hundred percent coverage. The requisite three sweeps of the six-mile by six-mile grid sectors were checked off the search map and no trace of Cessna N9388A was noted. Several ground teams had investigated local reports of downed or low flying airplanes and found nothing.

The official search lay entirely in the hands of civilian volunteers now, but the Oregon Air Force Reserve unit kept up its training regime, taking Al Jr. along as an observer. Between the Reserve unit and the CAP there was no coordination; each responded to their respective marching orders.

As happens during a search when time passes with no promising results, the searchers thought the crash had destroyed the plane and the human remains were covered with snow. Surely, if anyone had survived, they'd signal like crazy. A downed airman could do a lot to be spotted, like burn fuel, brush, or tires. They could cut trees and lay branches out in open areas; they could wave clothing. The trick to being seen from the air was to provide a contrast against the ground that could be spotted from above. Some people thought time would solve the disappearance, like it usually did in the mountain country of the borders. Usually as snow thawed in the warm spring air, they re-appeared. After the unsuccessful search for Art Ward in December, the searchers decided to reconvene in April when the snows melted. Surely they'd find some evidence of his missing *CallAir.*

Al Jr. had a different take on the circumstances than most: if the area was really covered one hundred percent, they'd have found the plane. And he knew how tough and smart Al Sr. and Phyllis could be when faced with a challenge. He knew they wouldn't give up without a fight or die easily. The son knew his father's family still existed, alive somewhere in the maze of peaks. Finding them was like searching for the proverbial needle in the haystack, but he'd give it his best and look through the mountains and trees until it made absolutely no sense to search further. He wouldn't accept "one hundred percent coverage" of any area until the mystery of his family's disappearance was solved.

Common sense and radar data indicated The Viking did not fly too far across the California border considering the March 11 icing conditions. Or could it have somehow exceeded the limits of the radar icing envelope and flown farther south than imaginable (or probable)? The Old Man had a knack for going further than most. After all, here was a man who'd parlayed a shoebox full of belongings into ownership of several hotels and businesses. Had he gone too far on this flight, somehow making it through the supposedly impenetrable ice-filled clouds and then crash-landed further south than the areas of the concentrated search? Had he turned east inland toward clear weather and made it to the vast Nevada desert? Had he gone west toward the coast or the ocean? Could there have been a navigation problem, an electrical failure leading to inflight confusion? Had he tried to go back to Portland? If he'd turned around to the north, surely he'd have called Flight Service in Medford? Who really knew?

In the late '60s aerial searchers used the "eyeball method" to find lost aircraft or people. For this system to work, the CAP loaded up a small general aviation aircraft with trained observers. These scanners looked for aircraft crash signs while the pilot navigated his craft carefully over predetermined six by six mile grids or flew other specific flight search patterns determined to fully maximize capabilities.

The observers strained their eyes looking for signs of a crash against the spotty background of snow, trees and rock. Square by square they scanned the sectors. Glints from metallic wreckage, broken trees tops, and burn marks could all indicate crash sites. If the missing people remained healthy enough to signal, sometimes the spotters saw their signs such as signal fires or SOS messages spelled with brush or stamped into the snow.

Emergency Radio, Then and Now

These days, with a radio signal emanating from a crash site, an overflying aircraft just needs to be tuned to the emergency frequency 121.5 or 406 MHz. Once locked onto a signal, it becomes a straightforward matter for a pilot to home in on the transmitter's location and find the downed aircraft by listening for the radio signals to become stronger as the plane flies closer to the wreck. If the pilot can't home in on a signal due to fuel or other issues, the general location of the signal gives rescuers a head start on the probable area of the crashed plane. Satellite discovery of emergency radio signals makes pinpointing the crash site even easier.

Emergency radios in 1967 were bulky, usually too big for small civilian aircraft. Though they'd been in existence since radios and airplanes were combined, there was no standard emergency beacon for light aircraft. There'd been the Gibson Girl radios of World War II, so called because their hourglass shape resembled the 1890s drawing of the famous Gibson girls. With the Gibson Girl radios, the downed aviator knelt over and straddled the radio, holding it between the knees. Then, cranking while transmitting, the airman sent his message to the world, either through voice or automatic Morse code signaling.

In Vietnam, military pilots carried bulky locator radios on their survival vests. Once down, the transmitter was turned on, sending out a signal to rescuers. With the Oien search, no emergency radio existed. The search remained strictly the old-time eyeball search, at least within the area of probability.

In a patchwork way, through the sporadic cloud conditions, the entire route from Medford to San Francisco got covered by airborne searchers, first by the noisy HU-16s out of Hamilton in the first hours after the plane went missing, then supposedly by a twin-engine USAF Beechcraft C-45 during daylight hours.

The California CAP Wing squadrons launched their daily flights from the Medford, Chico, and Yreka airports. No matter how hard people looked, the whereabouts of Cessna N9388A and her occupants remained a big question mark.

This fourth freezing day ended on Shoemaker Bally as another drenching storm tentacle unwound and slung toward the coast from the offshore low like a slow moving whip. For the Golden State's residents, the weather bureau's travel warnings had become routine, almost like that of the boy who cried wolf. Precipitation smashed existing records. Travel advisories for strong winds and heavy snow warned motorists and pilots to stay put and the Oiens finished the day just as they'd started it; huddled together in their cold prison; doing their best to stay sane and warm.

DAY 5

Wednesday, March 15, 1967
   Diary:

"Altimeter dropping. Staying in."

By now the daily routine had been set for life in The Viking. The best actions for crash survival were completed at the start of the ordeal. Rationing, shelter, warmth, injury prevention and treatment were done. Staying in was the only thing that made sense. All they could really do was conserve warmth and energy and wait for better conditions outside. Once they could get a good look at the surrounding country, maybe some decisions and planning could be made. So far the best walkabout had been made by Carla, the day she lost her bearings for hours, stumbling around in the blizzard. Her blundering back upon the wreck was mostly a matter of luck.

The list of priorities in an aircraft accident is generally this: Take care of life threatening injuries. Phyllis had done all she could to stop the flow of blood and she'd splinted any broken limbs. Preserve body core temperature by securing shelter from the cold, using whatever clothing, supplies, etc. are at hand. If necessary, create a shelter away from the crash but near enough to be spotted until the wreckage is safe from fire or explosion. (Contrary to Hollywood movie depictions, air crashes usually do not explode.) Use airplane parts to assist in survival. Fuel, batteries, wiring, insulation, etc. can all be used to make shelter, signals and heat. Take every advantage to make signals by fire, smoke, reflective items, etc. Lastly: "Hunker down and conserve energy. Do your best to keep faith that a rescue is on the way. Employ positive conversations to keep spirits up. Help is on the way, but you're going to need to be patient, especially if the weather is bad enough to slow down the process." *(From the blog, wilderness-urban-survival.blogspot.com)*

Hunkering down.... Since Saturday's fiasco all they'd done was safeguard their energy through the abysmal days. They'd followed the correct procedures for post-crash actions and lived through snow and cold and the terribly powerful, biting wind. To venture out from the wreck meant certain death. Staying in preserved their lives.

The altimeter in The Viking survived the crash impact and, like a barometer on a ship, provided a useful tool for predicting changes in the atmosphere. Noting the drop in air pressure, the three expected the snow, fog, and cold only to get worse. Carla's bleak report on the apparently uninhabited area convinced everyone that the best course was the one they followed. As they rested, they wondered why their rescue was taking so long.

If they could have read the nearby San Francisco newspapers, they'd have read about Senator Thomas Dodd embezzling money, a youth throwing himself in front of the president's limousine, and the class of 1967 graduating into a nation without jobs. They'd have noticed how their disappearance was fading from the news already, although they'd only been gone less than one week. If they'd seen how fast they became old news, they'd have felt abandoned!

Checking the weather section, they'd have noticed an article, "Calm Before the Storm," warning of an impending storm about to crash onto the coast one more time. The previous day along the nearby Sacramento Valley there'd been a respite from the rainy, windy, wet March, but the Trinity Mountains got no reprieve once again. In the Trinities the climate was only bad or worse.

An evil gift that kept on giving, the persistent low pressure cell lying off the Washington coast spawned round after round of wind and rain, with the worst onslaught aiming right for California-Oregon border country once more. Challenging weather was nothing new to the area. Historically the region had a reputation for sloppy weather, a fact utilized to the advantage of the U.S. military during World War II. Not too far west of the Oien crash sat the town of Arcata, where, prior to World War II, the U.S. Navy built an airbase, assuming the proximity to the Pacific Ocean would be useful for launching coastal patrols.

Unfortunately the thick coastal fogs prevented the base ever from reaching full utilization. The Navy discovered Arcata was the third foggiest location in the country. Later in the war the Arcata base became one of three U.S. bases used to develop and prove a special cloud burning technology. The equipment and techniques helped U.S. bombers operate to and from the soggy English airfields on their European bomb runs.

An evil gift that kept on giving, the persistent low pressure cell lying off the Washington coast spawned round after round of wind and rain, with the worst onslaught aiming right for California-Oregon border country once more.

In spring, a majority of the Arcata days provided IFR (instrument flight rules) conditions in fog, giving ample opportunity for testing. The special fog burners heated a mixture of diesel oil and aviation fuel and were built alongside the Arcata runway. Once lit, they burned the fog away using a tremendous amount of fuel, but allowing an aircraft to safely find the runway centerline in zero-zero flying conditions.

This technology was tested at other places too, the sort that experienced the horrible, wet, weather of Northern California. Amchitka Island in the Aleutians and Whidbey Island Naval Air Station in Puget Sound were used to prove the system.

Some airlines utilized the leftover technology after the war while flying scheduled service to Arcata. If the weather conditions were reported to be below landing minimums, the captain requested the control tower for the runway flames to be turned up. The tower lit the remote controlled burners and the burning fuel mixture cleared the fog. The DC–3s landed in legal weather conditions, much to the surprise of the passengers, who noticed the flames leaping fifteen feet high on each side of the runway. If only the fog burners were stationed on the side of Shoemaker Bally Peak, maybe the overflying planes could see something besides white clouds and glimpses of rocks!

DAY 6

Thursday, March 16, 1967
Diary:

"Opened plum jam and drank water from trap. Carla tried again to get out. Snow too deep and wet and freezing. Al started out after she returned. Sounds like planes flying. 0115 Al shouted across valley 'OK' on way for help. Only nice day so far."

Since the accident, the mountaintops remained covered up under obscurations of cloud. With temperatures consistently hovering below freezing, along with the constant snow and wind, there'd been no chance for them to survey their surroundings until this day, the "only nice day so far." As it has for eons, Shoemaker Bally stands tall within sight of Mt. Shasta. The north side provides an impressive panorama of rugged peaks and deep valleys. Even with a device designed to travel across the snow, the Trinities don't allow any ease of access. To the locals, the mountains shut down in the winter, put off-limits by nature.

To three people facing the specter of starvation, frostbite, gangrene, and hypothermia, the possibility of having to walk away was akin to facing a suicide mission. How could anyone cross the miles of snowy, brushy land, when just walking a few yards from the plane to urinate took such an effort? How could anyone survive a night in the open, in the deadly mountain air, when the three barely made it through the nights in layers of clothing combining their warmth in the cramped fuselage?

Unless they had snowshoes, sleeping bags, food, packs, and were in outstanding physical condition, any attempt to walk out would be a last ditch effort. Whoever left the safety of the plane took a heroic gamble. There could be no second acts; leaving was a one shot deal. But waiting with the plane only made sense if search and rescue was being done, and so far, it appeared no one was really looking for them. The high-flying airliners obviously were not, and the handful of planes buzzing at the lower elevations were not flying search patterns; they only passed through.

For almost a week, Al, Phyllis and Carla had been greatly resourceful with very limited supplies, doing what they could to keep breathing another day. As if following a survival text, they'd done everything possible to keep alive. Already they'd outlived the span of days usually allotted to mountain crash survivors, and they were aware their days were numbered unless help arrived. There'd been nothing but falling snow since the accident, so keeping the plane clear and clean was impossible—not that anyone would have seen it through the overcasts. The plane disappeared further beneath the white blanket with each snowfall.

Carla was in the best condition of the three, although her feet suffered terribly from the barefoot return walk a few days before. At fifteen she had youthful stamina and could possibly still walk out to somewhere. Her feet now showed signs of permanent frostbite damage and would not last much longer. If they deteriorated further, she'd be unable to walk even a short distance. She had to get moving before this happened. On this day she set out for her second reconnaissance, leaving the safety of the plane with the hopes of finding a way out of the nightmare. By now her feet were so swollen she clambered out in Phyllis's adult-sized shoes.

It took fortitude for a young girl to set out in the mountains with no food, gear, or even a clear idea of where to go, but Carla set out once again with high hopes. Without any real food for nearly a week, her stomach protested and grumbled as she exerted herself. It felt exhilarating to get outside after lying in the wreck for a week. The high country air chilled her to the bone as she crossed the vine-covered latticework that had grabbed the shoes from her feet just days prior. Walking on tightropes coated with ice, combined with the pull of gravity and the steepness of the slope, it was a feat worthy of a Wallenda to make any progress at all. Carla fell frequently, sprawling face first into the snow as her feet and legs poked through the layers of brush. In her letter about her first venture out, the one when she lost her shoes, she wrote to her friend about her hopes to find a logging road. If she could get to a road, the going would be much easier and safer. Roads led to people. Back in the plane, the meager food supply, now down to just a spoon of jam daily, really did nothing to fortify them. People in concentration camps received more nutrition. The M&M's had long been finished at the family's rate of 18 per day, six each. They stayed hydrated using the "water-works" bathing cap rig. Fortunately they had extra clothing, thick sweaters, and a blanket to help ward off the chill.

While Carla floundered on the hillside, Al and Phyllis likely spoke about the reality of their situation. Clearly with no food, injuries, and the lack of search planes, someone needed to walk out to find help or they'd die. While they both had a fairly clear idea of their location along the airway, no one could pinpoint the position. This was due to the fact that the last part of their flight occurred in clouds in an ice-covered airplane and the descent covered several miles. Their best guesses were miles off. Al was familiar enough with the flight route, having flown it many times since 1945. He knew the mountains were criss-crossed with logging roads and knew they were close to Redding and the Sacramento River Valley, that four-hundred-mile long river valley which ran north and south along California, sandwiched between the coastal mountains to the west and the Sierra Nevada Range to the east. The time had come to take some drastic action. Something had to be done.

CHAPTER 8

Al's Lonely Walk

WAITING AROUND FOR A savior grated against Al's nature to solve his own troubles. The cuts, broken bones, and blasting snow had forced him to hunker down since the accident. He'd always made his own luck by going out and getting dirty, whether it was crawling in the mud to fix one of his dump trucks or punching someone who tried to tell him how to run the hotel. If he'd been the type to wait for things to happen, he'd still be working on the ranch back in North Dakota with his brother Percy. The days of waiting since the crash weighed brutally on his mind, giving him a big headache, worse than the one he got from slamming the instrument panel with his face. The lack of search aircraft flying overhead meant no one knew where The Viking crashed.

He formulated a plan and decided he'd make a break for safety once the snow stopped. Right now it looked as if the plane would end up buried by the drifts. At least the snow could help by better trapping their warmth in the fuselage. Travel across the terrain would be rough in deep snow, but he knew he must try. He had a general idea of the crash location and figured he'd find help not far away.

They'd almost made it to Red Bluff. His authority as pilot-in-command made him completely responsible for the accident. He'd been chummed along past each checkpoint and felt slightly deceived. If the Beech Bonanza pilot had not reported the tops over Redding at 10,000 feet, he'd never have flown past Medford. Maybe the Bonanza pilot was correct, but where Al expected the clouds to level at 10,000 feet, they blocked the actual flight path, forcing him to plough right into instrument flight conditions—something he was not rated to do as a pilot and he'd diligently avoided so far on this trip. After years of safe flying, Al finally got bit in the worst way imaginable. Had they really been so lucky the day they lived through the accident? Maybe it would have been better to have slammed hard

*After years of safe
flying, Al finally got
bit in the worst way
imaginable.*

into a granite wall and gotten it all over
with at once.

Carla's report from her first venture
out on Monday hardly made him jump for
joy. The immediate surroundings sounded
wild. The "experts" always said to stay with
the plane, but in this case, they'd all waited long enough. Where the
hell were the searchers? No planes had flown overhead in the past
days. He'd always dropped whatever was going on in his life to help
search for missing planes with the OPA. Where were all the planes
when he needed help? The Feds always said to file a flight plan just
in case something happened, and he'd done so.

As The Boss he'd always been the lead dog and he knew it was
his duty to save his wife and daughter. Besides, with no medical
treatment, his broken bones would not heal right. It felt like a buzz-
saw slashing at every cell in his body whenever he moved. Phyllis
patched him up with rags, but it did not help much. Each breath
painfully ripped into him as broken ribs grated against his sternum
bones. His head ached from slamming forward in the crash and he
was weak from blood loss as a result of his cut face. His back hurt
bad, right where he'd broken it years before in the Chehalis crash.
Maybe he shouldn't have cut himself out of that body cast back
then, but at the time the Cuba flight with the club seemed like a lot
more fun than lying around rotting in a lousy hospital—a place he'd
be most grateful to be right now. He was much younger and more
resilient in those Chehalis days, he reckoned.

Phyllis knew her husband could die while venturing out of the
plane's shell, but she faced reality: They all could die if help did
not come soon. Of the three, Al was best suited to go. He'd been a
woodsman and knew how to find his way without a compass. Carla
had panicked the time she left the safety of the plane, and her feet
developed serious frostbite. She'd never make it alone outside the
plane, though if anyone were to live, Phyllis wanted it to be Carla.
She'd gladly trade her life and that of Al's so Carla could live. At
only fifteen years old, she deserved a longer life.

Their seeming abandonment took Phyllis's breath away. How
could something like this happen in modern day California? Did
the state not have a search and rescue outfit? What about the flight
plan and the position reports? Phyllis's feet were already hopelessly

destroyed from the freezing suffered that first night when she was delirious and didn't keep them covered. The skin was already blackening from dying tissue, and they had little feeling left in them. No way could she hobble far on them now, even if the place suddenly turned into a tropical paradise. They'd need to be found to survive. Like Carla, she pinned all their hopes on Al. He'd always been a good provider and was as reliable as a Swiss watch, doing just what he said he'd do.

Over the years, the Trinity Mountains provided huge harvests of timber and gold. Just a few years before 1967, loggers pulled a million board feet of lumber off The Viking's mountainside grave. The gold mine closed in the early 1900s after a thousand ounces of gold got yanked from the mountains' dirt. There were still plenty of signs of commerce to the naked eye, if the snow ever quit. The old mine road passed a few hundred yards below The Viking's gravesite. A logging road passed not far downhill from the crash site, too. Roads led to people and help. The snow-covered roads criss-crossed the country like white spidery veins snaking over the green mountains. They were impassable in the winter, but in the warm months meat-hungry hunters drove carefully along these pot-holed dirt roads hoping to blast an unlucky black-tailed deer. For the California DOT, U.S. Highway 299 was hard enough to keep clear in the winter; the mountain roads were completely ignored by snow removal crews.

To the people who'd been stuffed like sardines and frozen in a downed plane for days, this first bright day gave relief. Although still damned cold, at least the sun shone down and the wind calmed enough that a person could look outside and not face a ground blizzard or bitter gale. Fifty miles away they could see Mt. Shasta, tall and white, standing alone on the horizon. To their north they saw a formidable blockade of mountains. Behind them they faced the impossibly steep slope of Shoemaker Bally.

While the pressure rose and the day was looking fine, Al decided to wait no longer to leave the safety of the plane. There was little to lose, except his life, and all to gain if he walked to help. If the Cessna got spotted by a search plane in his absence, Phyllis and Carla would at least point rescuers in his direction. Like the plum jam, Al disappeared, slipping out through the jagged hole of the right side windshield. As his namesake son told the *Saturday Evening Post*, "It took guts for the old man, injured as he was, to walk away and leave his family there, leave the little safety that the plane afforded."

Alvin F. Oien, Sr. personified guts. Fifty-nine years old with an injured back, debilitating bruised ribs, cuts, contusions, no food, and a broken arm to boot, he dug deep for the tough Viking spirit whose roots had carried him so far. A smart man and not one prone to making rash choices, he brainstormed this one to the end. He'd walk out even if leaving shelter went against the opinion of the experts. Before he left, he insisted that Phyllis and Carla stay with the plane no matter what. They'd discussed their situation over and over for days.

Jack Jefford, the well-known Alaska pilot, learned full well the survivor's dilemma of staying with the plane or walking out after he crashed a Stinson in the Darby Mountains between Golovin and Elim, Alaska, in the late 1930s during a November storm. Jefford's plane came down with very little damage when it was caught in a mountain rotor wave, landing in a snowbank with minimal forward motion. In his book, *Winging It!*, Jefford described the situation eloquently after surviving for several days in a blizzard with no search aircraft looking for him:

> "The hardest thing was just passing the time… Sleeping was my only escape. By the third day I found that I could almost hibernate and I increased my sleeping period to nearly twenty hours.
>
> "A sense of hopelessness came over me. I tried to give myself pep talks, but logic told me I might not survive. Each passing hour brought me closer to that realization. My mind began to wander and all sorts of grim thoughts kept cropping up. By the end of the fourth day I was totally discouraged, realizing it was possible I'd never be found. Now for the first time, I wondered if I should try to walk out. There are few options open to a pilot who survives a wintertime crash in Alaska's wilderness. As in my situation, you are usually reduced to two grim choices: stay with the plane, hoping searchers will rescue you before you die; or strike out for civilization, gambling you'll find food and shelter before being overcome by exhaustion and the elements. Down through the years of Alaskan aviation history, scores of planes have vanished without a trace. Undoubtedly, among these

tragedies have been instances of pilots who survived and had to choose one of these terrible options. As the days dragged by I was continually faced with the same dilemma. Walk out?"

Ultimately Jefford made the decision to stay with his aircraft and by the sixth day, an over-flying Ferguson Airways plane found him.

(Note: this quote is taken from *Winging It!*, copyright 2009 Jack Jefford, Alaska Northwest Books, and is reprinted with permission.)

As a side note, the difficulty of finding a plane against snow is illustrated by Jefford's incident. The distance between Golovin and Elim is less than twenty miles. The terrain is rolling, with few trees. Jefford's route was well known, basically a straight line. In spite of these advantageous factors, finding him proved a very difficult task for the searchers. Survival experts who say that air crash survivors should stay with their aircraft don't always consider the situation faced by the longer-term survivors, the ones who have followed the rules and stayed put and who decide or realize no one is looking for them—the ones who lived through the window of survival possibility. How can anyone really understand the dilemma unless they are in the situation?

What if your plane is lying on a remote peak, buried under snow, and you are badly hurt, out of food, and have been there for a week with no search planes flying overhead? What is there left to lose as you're becoming weaker by the day, watching your family deteriorate before your eyes, knowing you helped cause the distress by making a mistake while flying? Should you still stay with the aircraft and rot until you cannot walk, or should you take your life into your hands and try to walk out for help?

Helen Klaben and Ralph Flores survived a March 1963 Yukon Territory crash in a Howard DGA (related in Klaben's story, *Hey I'm Alive*) and spent over a month in cold temperatures that rattled their bones. Both stayed in the crashed plane, hoping they'd be found. Finally, when it became clear no searchers even looked for them anymore, the two left the shelter of the wrecked Howard and ventured out for help. For them it paid off. Flores set Klaben up in a camp less hidden by trees than the plane, then walked toward a sound they'd heard far off in the hills. Within a day a bush pilot spotted Flores' SOS signal pointing to the injured Klaben's camp. If they'd not left the aircraft, they'd still be in the in the Yukon hills.

According to the Oien crash diary, Al waved to Carla just as she returned to the plane from her second trek, where she found herself floundering in armpit deep snow again. Shouting and signaling to her, it is likely he'd been assuring her they'd be okay. She only heard the "okay" part.

Back at the plane, Phyllis tore their blue-and-white checked blanket into strips and bandaged her daughter's feet. After her second mountain recon, Carla's feet had ballooned up even larger, so much that her mother's size ten shoes now fit her perfectly.

While his daughter recovered from her long walk, Al toughed it out, working his way downhill through the deep snow. The warm sun of "the first nice day" settled behind the peaks to the west, laying shadows across the valley. With the setting sun, the chill returned to the air. Al's labored breath came out in clouds of frosty air. After almost a week in the cramped plane, he'd be on his own for the night. He was cold, although he'd put on just about everything he owned. As Carla's notes stated, they all wore up to seven layers of clothing.

To pilots, a "sucker hole" is a break of better weather during a span of lousy weather. The pilot paces on the ground, biting his nails, waiting for the foul conditions to pass so he can launch. When the sky clears enough to take off, he ventures into the air, only to find that the "hole" in the weather is actually just temporary and he's forced to fly in the soup—suckered in, as they say. If he is lucky, he can get back on the ground safely; if not, he joins the long list of dead pilots lost forever to a sucker hole. The altimeter needle rising toward a higher pressure indicated only a short-term improvement in weather for Al. As fast as the needle peaked, it slammed down again, registering the lowering pressure of the incoming storm wave. The "first nice day" was a sucker hole in a span of stormy days.

A Northwest outdoorsman all his life, Al likely owned the best outdoor gear, but it all remained back in Portland. No one had planned a stop in the Trinity Mountains on this flight. A week ago he'd envisioned himself sunning in Texas, discussing financial matters with his eldest son, getting to know him better. What a far cry from the Texas sun this godforsaken mountainside hellhole was, he may have thought as he floundered through the armpit-deep snow. His financial troubles seemed wonderful compared to the misery of present.

As the Boss stumbled down the snow-covered mountain, torturing his lungs with the exertion in the cold air, fog settled in the lower parts of the valley and the air pressure dropped with the settling weather. Clear days did not last for long in March in the Trinities. Fog brought drizzle, and he knew he'd be facing a rough night in the open country.

He'd fallen for a sucker hole, the sort of trap he'd avoided all his flying life. There was no way he'd make it back uphill to the plane, so he'd best keep moving onward. There were logging roads all over the place in these treacherous mountains and all the roads led to civilization. Finding the right one through miles of twists and turns would be the trick.

He'd walk until he found something or someone who could render aid, or he'd die trying. The sunlight faded. The high peaks may have echoed teasing sounds of high-flying jets. He shuddered from the cold and his teeth chattered. He pulled his city clothes to him but they did nothing to help keep him warm. His body was soaked with sweat and he was freezing at the same time. Dangerously cold now, the lost heat could not be replaced, and hypothermia attacked.

Alvin F. Oien, Sr. knew he must keep moving just to stay alive. Breathing with broken ribs was painful as his intercostal and chest muscles heaved in the thin mountain air. Each steamy breath felt like knives plunging in and out of his lungs. The broken arm and wrenched back sent jolts of pain throughout his being as if he'd been electrocuted. He focused on mind over matter, blocking the sensations, and resolutely moved downhill, sprawling and slipping, sometimes sinking and plunging thigh-, then armpit-deep into the snow over the latticework of snow-covered vines. His mouth was parched from thirst, his body soaked with sweat, and his heart raced from the efforts. To top it off he was starving and weak; although the hunger pangs bothered his stomach less and less. His head spun from dizziness.

There was nothing for Al to do but eat snow, rest a bit, and walk on through the night. If he came upon a logging road, maybe he could walk in the darkness and not worry about plunging so deep into the undergrowth. If he did get to a road, maybe he'd shelter under a log or tree, and then move in the light of day. He'd spent many a night out in the woods, just not while in such a broken condition. He missed the shelter of the fuselage and the body heat from Phyllis and Carla. He missed his life in Portland.

Positive force is what kept Al putting one foot in front of the other, though he suffered from the type of injuries where he'd be hospitalized in critical condition were he near a hospital. The belief he'd find a rescuer for Carla and Phyllis kept him moving forward step by step in spite of the excruciating pain. Always an optimist, on this walk he even wore his Optimist Club pin.

Maybe some doubt crept into Alvin F. Oien's mind, and for the first time the final end loomed on the horizon. He'd always bludgeoned his way through life, the only way he knew how, taking out obstacles as they sprung in his path, never afraid, never giving up, always finding a way. He'd punched, fought, and scratched his way along and not once had he feared the end. Was the game up? It was all up to him.

In the same diminishing light and cooling air that chilled the lone walking man, the CAP search planes returned to their airport roosts after flying all day and covering the patchwork of square miles. Of course, there was no sighting of The Viking. No one even came close to the wreck on this "first nice day." The searchers still flew a hundred miles away to the north. Maybe the mission could wrap up in the coming days. The CAP geared up on the weekends when more people escaped work to fly the planes and look out the windows. Their efforts slacked slightly during the weekdays when the volunteers reported to their regular jobs. This Saturday and Sunday, there'd be a big push to maximize eyeballs and airplanes. Maybe they'd find the plane with this maximum effort mission. The weather map finally showed promising conditions, and if it proved to be accurate, the planes could cover ground and make up for lost time.

No one even came close to the wreck on this "first nice day." The searchers still flew a hundred miles away to the north.

Over the past week the atmosphere had coughed up some of the worst weather in California history. Somewhere a family lay in the wilderness. In their hearts, the search pilots and professional rescuers knew no one could survive in those conditions, but as long as the CAP mission leaders elected to keep the search alive, the loyal volunteers would keep flying their planes flying and eyes open.

While the Boss floundered down the mountain, fighting the first stages of hypothermia, hundreds of miles to the north his oldest son flew over the country, viewing it through the side Plexiglas blisters of the Oregon Air National Guard Grummans. Each night they flew back to base in Portland, landing at the International Airport that the Boss helped to build, delivering the gravel with his dump trucks. The Oregon Air Guard HU-16s would not fly training missions forever as a favor, and when they were done he'd rent his own search plane. Like the Old Man, Junior would not give up. He vowed to scour every square inch of the country between Portland and San Francisco with his eagle eyes until he found the plane. Junior planned to search until he ran out of time or money, or life. Junior had no idea his search flying would soon almost cost him his own life, and how, through a close call, he would find out what happened with his father's plane.

While Al Jr. took care of the search-flying, Ron and Chuck continued to run the Clifford Hotel, keeping the doors open and the bills paid. Running the Old Man's affairs was not part of their life plan but they knew it must be done. Though they'd been estranged, they knew someone had to fill the Old Man's shoes. Of course he was not dead. He'd make it through and come walking out of the woods, and no doubt he'd give those search pilots a piece of his mind. It was just a matter of time.

Back at The Viking, Phyllis cuddled with Carla, conserving their pitiful warmth against the frigid air. They wrote how they missed Al's heat but liked having more room in the plane after he left. Phyllis tended the girl's disintegrating feet and her own blackening ones. She realized the distinct possibility they'd not survive unless help came soon. She knew she must stay strong for Carla. She kept her fears to herself. Fear spread.

The two just had to stay put and keep hoping someone saw the aircraft or that Al found help. If help did not arrive soon, they'd both lose their feet at the very least. She knew Al was in bad shape when he left too, but she also knew of his immense inner strength. He'd come through or die trying. She could feel it. They'd talked before he left and both knew he'd taken a long shot and played his last card. Like the last flight, there'd be no turning back. He'd return with help or not at all.

CHAPTER 9

Week Two

DAY 7

F RIDAY, MARCH 17, 1967; Diary:

"Fog, rain, Carla went to creek for water, heard shouts about 215, sounded like copter near but saw nothing in the air. Cleared about 5. Phyllis period—ugh"

The Oien family spent a week under the unrelenting snowfall on the godforsaken mountain with an unknown name. Their standard of living fell from that of sleek suburbanites down to that of injured, dying animals.

With Al now gone, the mother and daughter focused on the simple tasks needed to endure through each day: collecting drinking water, updating the diary, resting, and staying alert for sounds of rescuers.

A drizzly fog had moved in over the mountains, once again locking The Viking under the impenetrable gray cap. At least the drizzle that came when Al left brought warmer temperatures compared to the snowy days, but still it remained too damned cold. The warmer air melted the snow and the fuselage became a dripping sieve with snowmelt and rain leaking through all the cracks. Staying dry was impossible, and the dampness brought a chill that cut to the core. Though warmer, the wetness brought more misery. They both anticipated help coming soon. And their hopes remained pinned on Al. He'd not been gone long, and he'd never let them down. Though he'd been rough to live with sometimes, he was very good to them over the years.

Maybe he found help nearby and a cavalry of rescuers was waiting for the rain and fog to clear before assaulting the mountain and saving them. Even if they could be spotted from above, their rescue in the remote, rugged country would be a difficult feat. Carla's two expeditions proved the impassibility of the countryside.

Since Al left, their ears and eyes were hyper-alert for any signs of other life. Other than an occasional bird or the high-flying airliners, they'd seen or heard no signs of any other life on the mountain. Surely these were their last days of suffering. Counting the facts in their favor convinced them they'd be rescued: The Viking was on a flight plan, using a route over well-traveled VOR airways, and military airbases lay all around them. Where were the rescuers, though? Didn't California have any search and rescue organizations?

Late in the afternoon they heard helicopter sounds and shouts, but they saw nothing in the gray sky. No helicopters or people were ever known to have come near the site and it's possible their minds tricked their ears into believing a rescue was at hand. Auditory hallucinations are not uncommon among starving, exhausted, or marooned people. Accounts of the lost frequently include rescue hallucinations where the stranded see and hear rescue planes, even responding wildly with waves and shouts of their own, only to realize the event was a cruel delusion. Or perhaps the shouts were actually Al's deep baritone voice, him trying to make his way back uphill to the fuselage, floundering back through the impossibly deep snow after a cold night of travel.

The events of the past week conspired to kill the family on all fronts but they remained unvanquished. So far they'd survived a plane crash. They suffered through cold, frostbite, injuries, and fatigue. They battled false hope. They fought starvation. The amount of calories ingested in The Viking since the accident was barely the amount required to keep a small animal alive. Now they ran on their body reserves, the naturally built in emergency supply most people carry—fat. Carla and Phyllis's bodies began to consume their fat, cell by cell. Each shiver or movement cost them another second of life. There was nothing to ingest other than snow, unless you counted a spoon of sugary jelly as food, and the jelly was almost gone.

If things did not change fast, they'd die on the mountain. Phyllis knew it as a fact. She'd always been a realist, and circumstances at the 4,300-foot level did not look as if they were changing for the better anytime soon. Carla remained naturally optimistic, still believing in her teenaged brain that a rescue would inevitably occur. As Carla imagined the inevitable rescue, another captive of a different sort,

James Stockdale, was formulating principles based upon the folly of optimism.

Captured over Vietnam when his aircraft was shot from the sky in 1965, Stockdale spent years in Vietnamese prisoner of war camps. Interestingly enough, afterwards when asked about survival traits and which prisoners were the first to fold, he related that it was the optimists who would say, "We're going to be out by Christmas..." then later "...by Easter... Thanksgiving..."—and each optimistic date would come and then go. Stockdale said, "And they died of a broken heart... You must never confuse faith that you will prevail in the end—which you can never afford to lose—with the discipline to confront the most brutal facts of your current reality, whatever they might be."

In the end Stockdale credited his survival to Stoic philosophy.

Phyllis was stoical, on the surface. In her private letters she wrote how it tore at her heart to see the daughter she'd nurtured so carefully, the one so full of life, thrown into a deadly game of survival, but she tried to not let on outwardly how much pain it caused. There were times Phyllis could not hold back and the impending doom prevailed. Together the two spent hours crying. They "just had to get out of here," their notes said.

Though the search remained active, the story of the missing plane and family had completely faded from the western newspapers, replaced by other, more recent tragedies. Not far away, another air crash seriously injured a group of California socialites. The death toll of the war in Asia continued to climb as American boys were killed or injured daily.

The Bay Area newspapers urged residents to put their umbrellas away, confidently stating that the storms which plagued the state for the past weeks were finished. The offshore low that blasted the West Coast for over a week finally dissipated. For the near future, high temperatures in the 70s would keep things drier. Maybe spring finally arrived in Northern California. In the Trinities, it was not yet spring. The Trinity Mountains still made their own weather, under their own terms. Even if the calendar said it was close to spring, the calendar meant nothing to the silent, forbidding peaks.

DAY 8

Saturday, March 18, 1967

Diary:

"SAT. Cold nite, looks like fog will burn off to clear."

Their priorities had changed drastically. Now they prayed for enough headroom to be able to sit up, for warm beds and food. They just wanted the fog to clear so planes could find them. Late in the day it did clear again and all the planes were looking hard for them, just in the wrong places.

With more volunteer CAP searchers out on the weekend, Saturday turned out to be one of the most active days of the air search, with eighteen planes launched into clear skies from the Yreka, Chico, and Medford Airports. Still following the interpretations of the radar and icing data collected the previous weekend, these planes flew all over the "area of probability," far to the north of the rough Trinity Mountains.

If unfolded and laid flat, the nooks, crannies, peaks, and valleys of Northern California and Southern Oregon would have added thousands of square miles to the search area. Because of the past week's weather conditions, anyone who might have survived a plane crash would likely have succumbed by now. Even in the warmer, populated lowlands, the severe weather killed people. Expecting anyone to live in the mountains through the storms was like expecting a miracle. Once in a while miracles happened, but not often enough for anyone to develop true faith in them.

Like paramedics who continue to administer CPR to a victim although they know death has occurred, the search pilots and their observers kept true to their mission and flew the assigned flights, checking off the squares on the map. Per CAP protocol, after three sweeps each sector could be cleared as one hundred percent covered, and the pilots flew on to the next square. With 18 planes flying all day, hundreds of square miles were cleared.

The Sunday aviation forecast showed a promising search day also. Over 25 planes were scheduled to fly all day throughout Northern California. The ungainly Albatross of the Portland Air Guard unit kept aloft on training missions with Al Jr. aboard. Each morning the deafening 1,400 horsepower Wright radial engines of the silver and yellow Grumman amphibians shook the buildings at Portland

International Airport as they revved up to full power during their long takeoff rolls. Thundering south along the same track as the missing Viking, the Air Guard search crews put in long hours craning their necks and wearing out their eyes over the valleys and gullies of the Siskyou country. After a long flying day, the big Grummans returned to Portland. The flight crew members knew Al's family was somewhere out there below the flight path. Perhaps they flew directly over the wreckage, but failed to spot it. That had happened more than once in search and rescue history.

SAR Techniques in the 1960s

Several techniques used in the 1960s to search for a downed aircraft are still used today, although GPS and other electronics have a bigger role in modern search and rescue missions. Famous Canadian SAR pilot R. J. Mokler explains the methods: The first type of search, called a "track crawl," occurs shortly after the aircraft is overdue. The search aircraft flies the route from the last known position of the missing plane to the intended destination. If the missing aircraft is not found, a parallel track crawl may be implemented where the search aircraft flies parallel to the route by a few miles.

During the track crawl, which is sometimes done at night, when overdue planes are usually confirmed missing, the airborne observers check for signs of distress, looking for signals such as bonfires or flares. Radio calls are transmitted on emergency and local frequencies, calling out to the missing craft while a listening watch is maintained over the airwaves. A track crawl is the sort of search the Hamilton-based Grummans did on the night of March 11, when The Viking missed its ETA into San Francisco.

A "creeping line ahead" search has been described as similar to polishing a row of tiles on a floor, where each square gets checked in an expanding pattern from the last known position (LKP) of the airplane.

The similar square or expanding square search is useful when the last position of the missing person or plane is definitely known, such as with The Viking. This type of search commences at the last known position and expands outward, square by square. This expanding search requires good visibility.

A contour search is a third method of checking mountainous

terrain for missing subjects. This method allows search planes to check uneven, mountainous terrain. The danger with contour searching is that a plane may fly into a box canyon, a place it could be unable to climb out from. Contour searching is an effective method of checking mountainous country, but very dangerous in lowering weather and in hills or high country.

Aircraft on preliminary searches try to fly approximately one thousand feet above the terrain over land and sea, keeping at least three miles of visibility on each side. Night searchers generally fly two thousand feet above land and sea. Altitudes vary during searches due to changing conditions, and nothing is ever the same for each situation.

The methods described were all used during the search for the elusive Viking. Ground search parties explore possible sightings also. Hundreds of people were involved in the SAR by this day; hundreds of hours had been flown now too. Hampering these great efforts were the fact that the devilish clouds made systematic searching very difficult. The pilots and observers did what they could among the rapidly changing conditions found in the mountain air.

Finding the gray colored Viking against the white snow was a worst-case scenario, a nightmare for searchers. The gray wings, even if they were cleared of snow, would remain invisible. Brush could have been used to mark the plane or lay out a signal, but without a saw, strength, and feet to walk upon, there was no way Phyllis or Carla could accomplish the task. Al's broken ribs, back, and arm prevented him from laying out a signal. Cutting the brush and pulling it along the mountain across the impossibly snarled vines would be difficult even for a well-fed person in perfect physical condition.

Finding the gray colored Viking against the white snow was a worst-case scenario, a nightmare for searchers

By Day 8 there was not much the mother and daughter now could do but wonder: Where was Al? When it got cold and foggy once again, could he even see or stay warm? Where were the search planes?

DAY 9

Sunday, March 19, 1967
   Diary:

*"Awful cold nite. Temperature 30 degrees at 9:30 AM. Fog
blowing off. Overcast. Carla for water. Lost most on the way back
by slipping. Have heard no planes. Our 9th day. One glass jelly
left. Getting weak. Fear Al did not make it to help. Making cards
out of upholstery to entertain ourselves. Feet need attention as well
as stomacks."*

For the first time in weeks, the sky cleared over most of the
West, allowing all of the search planes to fly in California and
Oregon. While people rejoiced in the warm blast of southerly air,
the marooned prisoners remained stuck in the ice box of Shoemaker
Bally Mountain. This barrier of mountains created its own microcli-
mate. Spring came a bit sooner on the southern slopes but they faced
north. The season's change would not occur until the sun shone
long enough to raise the temperature above freezing for more than
a few hours each day. This would not happen for another month or
more. Since the accident, not once had the temperature registered
much above 30.

As usual, on this Sunday the thick clouds clung to the peak above
the crash, blocking any chance of being spotted even if all 25 planes
filled with observers flew overhead. For the past week, conditions
varied little from snow with mixed rain to strong winds seasoned
with the drizzly, freezing fog. Aerial searching peaked on this day
with the 25 planes launching at one time, flying from morning
until nightfall, amassing nearly a hundred hours of flight time. As
the planes searched all the wrong places, the two women remained
hunkered down in the aluminum shell, waiting patiently as they'd
done for over a week. Al had told them to stay put and so they did,
there being no choice in the matter anyway.

In their notes they wrote how they both suspected Al had died
somewhere out in the valley below them. Phyllis knew his medical
problems were life-threatening without treatment, and his hike out
was a risky, last ditch effort at best. She'd seen many a patient in
his condition hospitalized, but definitely not out in the wilderness
trekking through the high country arctic conditions in city clothes,
injured and without anything to eat.

It must have been hard to believe the strong man she'd known, the one who once lifted her over his head and put her upside down in a snow bank, was gone for good. "The Boss" had lived with such strength and vitality and was such a big presence in their family life. His baritone voice echoed in their heads: his di-dee-do song, bossing them around, commenting in his succinct way about everyone and everything.

Surely he'd have led a rescue by now if he'd lived. Both mother and daughter remained haunted by the shouts heard the other day; maybe it was just Al shouting for help, the only time he'd ever needed help from them in his life. He'd always been the rescuer in one way or another, when he'd drop what he was doing to fly search-and-rescues when someone went down. There was even that time with the OPA, when he found the plane crash.

The two constantly fought off crippling emotions and depression. Depression paved the road to desolation, something to be avoided at all costs. As long as they kept breathing and kept occupied through the long days, a chance existed they would survive and be found. The mountainside and aircraft were now under snow so deep they'd never be able to get out of their predicament by walking on their own, even if their feet were uninjured. Just clambering through the windshield on the frostbitten limbs was a major and painful extrication. Carla continued her trips to the creeks to collect water when possible. The snow around the aircraft was dirty now and the running creek provided the only fresh water supply.

Time was filled with useful tasks. Cutting the small deck of playing cards from the silver Naugahyde upholstery took a few days. Each card got painstakingly trimmed into a small uniform rectangle. Then, with the red pen used to write the diary, each suit and number was drawn carefully onto the card. In the end they had their own deck of 52 cards, which they used for games of pinochle.

To pass more time they took turns reading their only book besides the Airman's Guide: *Fate Is The Hunter* by Earnest Gann. There could not have been a more depressing book to read. The 1962 bestseller featured story after harrowing story of Gann's career in commercial aviation in the 1930s and 1940s. A popular book that was turned into a John Wayne movie, it carried the premise that all surviving pilots were actually lucky. Those who lost their lives in aircraft were hunted down by fate.

A new future was planned without Al, and dreaming up long lists of food occupied their tortured minds. Maybe they composed their final letters in their head. Soon they'd begin writing those letters.

Together they became sicker and weaker from the systemic poisoning caused by their frost-damaged feet. Their skin tissue died off from the gangrene of frostbite. The lack of nutrition kept Phyllis's broken bones from knitting and the unceasing cold wore them down, forcing them to burn themselves away by constant shivering. Being depleted of glucose now, their brains worked only sluggishly while they tried to maintain their grip on life. Down to one glass of the plum jelly, they continued the rationing. Each daily spoonful jolted their memories just enough to remind them of the plum trees in the yard and of better times in Portland. The jelly just prolonged the agony.

On the snow-covered mountainside they still would not find anything to eat. There were small animals that burrowed beneath the snow and there were plants, but no way to dig down and capture or harvest them. When the jelly glass finally emptied, they'd begin their diet of snow only. Their bodies were dying before their own eyes. Gangrene affected their frost-damaged skin, causing their flesh to blacken where it had frozen on their feet. Protecting the skin helped, but proved to be losing battle without the clean and sterile dressings required to treat such a condition. If treated early enough, frostbitten and gangrenous skin tissues can be saved. In remote places where medical treatment is unavailable, sufferers of dry gangrene will sometime slough off their dead tissues or even digits in a process called auto amputation.

Earlier in the week, when Phyllis cut the blanket into strips and swaddled Carla's frostbitten feet, she'd done all she could medically. There were no salves or antibiotics or whirlpool baths like they'd use in a hospital. Covering Carla's feet with the strips was a just stop-gap measure to provide some protection until they got to a hospital. Compared to now, the previous week seemed to have been a better time. Luckily Carla's feet still worked well enough she could barely stumble down to the creek to carry water, but she'd definitely not be walking cross-country on them.

As the sun dipped and the evening cold set in, the search planes returned to their various airports scattered across the border country. Never again would so many planes be in the sky searching for the fallen Viking. Statistically there was a zero chance anyone who survived the crash remained alive. Though officially still open, the search for the missing family slowly unwound just as it had geared up over a week before. Over the next days, planes returned to their home bases one by one, pilots returned to their jobs, and the news completely disappeared from the media. While most searchers threw in the towel, Phyllis and Carla remained hopeful they'd be sighted. Al Jr. remained faithful to his quest and showed up each day at the Portland airbase, dutifully taking his seat on the SA-16.

DAY 10

Monday, March 20, 1967
    Diary:

"Rain, Snow. No Change. Feet painful. No trip to creek. Collected rainwater."

More monotonous weather kept them locked down in the aluminum tube for one more day. Being there was worse than being in prison. At least prisoners had food, shelter, and medical care. Shoemaker Bally never changed. The rain, snow, and constant fog steadily ground away at their nerves. Getting water from the trickle of a stream running to Grass Valley Creek became Carla's only reason to leave the fuselage home. Her feet hurt like hell, but when they needed water she hobbled to the creek. At least rain brought the water to them and she could rest. They collected what they could with whatever containers were on hand, like the empty jelly glasses, also collecting and squeezing water into fabric, then wringing it into their mouths.

As the official SAR entered week two, the scope of the search was already limited. It was generally believed that the aircraft lay somewhere west of Mount Shasta, but no one really knew except the survivors, who could see Shasta off in the distance to the northwest. Al Jr. kept up his airborne vigil, while Ron and Chuck Oien handled the hotel business. Various civilian pilots and CAP squadrons kept their aircraft flying, but no new information of The Viking developed.

Besides weather problems, other scenarios that may have affected the missing plane were considered. Maybe the aircraft suffered complete radio or navigational failure and continued into oblivion. A faulty compass could easily skew an aircraft onto a completely wrong track. With the speed and range of the Cessna, the area of its disappearance could be vast. Disregarding these possibilities, the main efforts still concentrated where they always had been: just south of Medford, Oregon.

Composite Squadron 34 returned to Eureka with its handful of aircraft, but flew missions staged from the Eureka Airport for the remainder of the week, checking from the coast inland. A chance existed that Al Oien's plane deviated west toward Eureka the day it disappeared.

The Oregon Pilots Association was responsible for air searching in the state of Oregon. On this day the Association announced through Director Bob Dunn it would assist in the search with five aircraft: two operating from Medford and three slated to check the steep walls of the Dunsmuir Canyon near where Art Ward disappeared in December. Searching the canyon made sense. It provided a low altitude route through Northern California and, if the weather lowered, canyons were the place VFR-rated pilots like Al flew to transit an area. The OPA search was centered on the Dunsmuir, California area, a region plagued by heavy snow, low ceilings, and high winds.

Prior to March 20, 163 sorties were flown for a total of 288 flight hours. There were 21 CAP planes in the air on March 18 and 25 CAP planes on March 20. The USAF had been using SA-16s, two twin-engine aircraft, and Portland's own 304th Rescue Squadron also participated. On March 20, Dunn received an official request for help from Oregon, specifically for five aircraft and competent pilots.

Ralph McGinnis was coordinating with Klamath Falls and Medford for these aircraft and their crews, composed of professional charter pilots familiar with the area.

According to the OPA newsletter of April 1967:

"As a matter of policy and preference, in the interest of overall SAFETY, local-regional pilots are used since their jobs have familiarized them with the terrain, weather, etc. and since they are conscientious and devoted. Naturally, their flight capabilities are current and often more proficient."

Friends of the missing are customarily rejected because of the emotional factors involved, creating the temptation to extend one's loyalty to a point of personal danger and resultant peril to the original rescue efforts. Dunn acknowledged the sincere offers of assistance from the chapter and assured members if additional effort was needed and the needs went unfilled, pilots would be contacted personally and immediately.

"High wing aircraft with extra horsepower are best suited and others are usually inadequate for search efforts. Dunn was unable to authorize any search in California, however has been in constant communication with the CAP and USAF activities there. If any dissatisfaction arises, he is in a position to communicate directly with the USAF."

Thanks went out to Norm Winningstad, Marv Watt, Bill Tingley, Ted Flaming, Earl Trigstad, George Marco, Walt Rupert, and all the others who participated in the Oien search, or attempted to participate, and all who volunteered their services.

Another section in the newsletter mentioned how the group missed their projectionist, Al, and the food coordinator, Phyllis.

DAY 11

Tuesday, March 21, 1967
   Diary:

   *"Overcast, Fog. Planes seem to be flying down our valley, tho to
   high to see us. Have lost all hope. Suggest Suicide pills in 1st aid
   kits. Can't try to walk out—snow too deep for us. Will stay here till
   gangrene, cold, or malnutrition gets us. Phyllis and Carla struggled
   to stream. Treacherous trip. Cold and wet on return. Exhausted too
   weak."*

Eleven days was a damned long time to be marooned without real food, warmth, sleep, or safety. These deficits took their emotional and physical toll on mother and daughter, and suicide was becoming a reasonable option with no end of the ordeal in sight. The past week was filled with enough grief to fill a lifetime. Since March 11, they'd crashed, Al was dead, and they were stuck on a snowy mountain without food or much hope of rescue. Clearly they'd lost everything.

Reminding them of their former and now unreachable world,

the high-flying airliners plodded along the coastal air routes flying north and south. There'd been no low flying search planes. The snow pack now covered the gray and blue aircraft, obscuring it completely from the air. Ironically, the deep snowpack still kept them warm by insulating The Viking from the mountain wind chill—the only benefit of being buried. The seven or more layers of clothes helped too, but barely.

They had covered the windshield hole with the canvas engine tarp, clearing it as they needed to exit. Moving in and out and putting the tarp on and off was laborious. Falling snow or rain was a daily event, making the tarp difficult to handle. As they weakened, it became unwieldy and almost impossible to move. Sometimes it froze to the plane and became as solid as if it was piece of steel bolted to the windshield frame. Later, as they weakened, it became too much to open the tarp or even get out.

When possible, Carla trekked to the creek and carried water back to the plane. Most of it spilled while crossing the impossible gap between the creek and plane. The snowpack had become much harder to transit over the past 12 days. It was deeper, wetter, and more unstable. Phyllis, because of her badly damaged feet, rarely ventured outside. Still, the two managed to collect enough water to stay hydrated. Experts say ten days is the maximum a human can live without water. Water from the snow and creek was the only commodity they possessed in volume. The meager spoonful of plum jam continued from the last jar, always reminding them, like the airliners did, of the distant world they'd known before.

They now thought of their old world in Portland the same way they'd thought of the distant stars: unreachable and a distant dream. After eleven days without real food, the human body functions but is beyond using its normal reserves and is actually eating itself by consuming fat or muscle, depending on the starving persons' physical condition, converting both to sustenance. Variables affecting the bodies' use of itself are obesity, climate, genetics, health, and hydration. A person in colder climates expends caloric energy just to stay warm. It takes 123 calories just to heat a gallon of freezing water to body temperature. The snow water diet and cold air helped them hydrate but they expended valuable body energy just digesting the liquid.

Body fat allows a margin of safety in a survival situation. Humans are genetically programmed to retain fat, likely designed to allow continual living through periods of famine. Women as a rule fare better than men in starvation occurrences due to their higher body mass index and lower muscle mass. Neither Phyllis nor Carla carried much excess body fat, something that could have helped them stay warm in addition to providing energy.

The brain is selfish during starvation; it takes from the body in order to keep functioning. In the first stage of starvation, glucose is provided by the glycogen stores, which get depleted rapidly. Next, the body fats are depleted. Finally protein and muscle tissue is catabolized to maintain energy and keep the brain synapses firing. Without food, their health problems compounded. Besides needing calories for injury repair and homeostasis, they needed vital nutrients. The human body can't produce vitamin C and must obtain a continual supply from outside the body.

Malnutrition victims develop, among a myriad of other diseases, scurvy, the curse of the early seafarers. Without fresh infusions of vitamin C from greens and citrus fruits, signs of scurvy can appear in as little as one week. The signs and symptoms depend upon the amount of vitamin C previously stored in the victims' body. Scurvy causes ailments such as loose teeth, dry hair, nausea, lethargy, body pain, and inadequate or slow healing of wounds; everything Carla and Phyllis suffered by day 11.

DAY 12

Wednesday, March 22, 1967
  Diary:

  "cold, windy, foggy, overcast. Heard a plane last night that could have seen us we hope. 1 day jelly left. water gone, snow now dirty. Feet worsening, lost considerable weight.

  foods most wanted.
  1 gal fr. or. jc.
  root beer
  whisky and 7up
  Mogen David
  coke
  milk

water to drink
soups
chili—crax. or bread and butter
cc dinner
pot roast
fruit—or. & apples.
baked pot.
grapes
diet rite and popcorn

pasty
fish & chips
oreos
Calif. crunchy cookies
Chinese food
macaroni and cheese
hot fudge sundae"

In Carla's handwriting:

"hot chocolate
hamburgers
French fries
ice cream- cones and shakes
ham and dill sandwich
ham
tacos—burrito
French dip sandwich
tuna fish
rice-a-roni

chicken
spaghetti
hot dogs
pancakes and sausage
bacon and eggs
toast and jam
English muffins
peanut butter and jelly
sauer kraut and sausage
peanut butter and celery

Both want
flat level warm bed
enough head space to sit up
service
relaxation—nerves shot
sleep"

The grueling days, the cold, the stress, and the daily challenge of staying alive chipped away at Carla and Phyllis. Typical of the hungry, they became obsessed with food and created their long list of dream foods, a typical reaction of those living with empty stomachs.

During a University of Minnesota study in 1944, conscientious objectors agreed to eat a starvation diet of less than 1,600 calories. Not surprisingly, as hunger pangs set in, the test subjects obsessed over food. Some wrote extensive recipes, some dreamt of food, and some made long lists of favorite foods, just as Phyllis and Carla wrote theirs.

The two already exhibited most of the symptoms of the mal-nourished university test subjects: lethargy, dizziness, and anxi-ety. Test subjects also experienced cold intolerance, edema, muscle soreness, hair loss, and some withdrew from their university classes because they'd lost the ability to concentrate. The test subjects' lim-ited 1,600-calorie diet was far more than the standard ration of one spoon of plum jelly aboard The Viking.

The prospect of slow decline from the synergistic effects of their multiple health problems became reality. Horrified, they could do nothing while their once-healthy bodies disintegrated daily. The dry gangrene ate at their blackened feet. The cold sapped their strength. Hunger gnawed at their stomachs. Grief and fear tortured them. Another factor slowly nibbling away at their sanity was sleep. Unable to eat, stay warm, or be comfortable, they could not even escape into deep sleep. Whenever they were able to drift off, wind, cold, or pain jolted them back to reality.

When deprived of normal sleep, test subjects begin to show signs of mental illness. Animals deprived of sleep during experiments have died. For Carla and Phyllis, quality sleep was nonexistent. The cold, the cramped space, the winds, the pain, the rumbling in their stom-ach, and their fears conspired against the two.

DAY 13

Thursday, March 23, 1967
 Diary:

> *"Wretched night. Poured again. Caught some water. Having
> stomach cramps and referred aches. Feet terribly swollen, toes purple.
> Reading "Fate is a Hunter." Snow, rain, hail, seen alternately all
> day with very strong winds. Heard two planes today but did not see.
> The 1125 explosion was much closer. Cabin is a dripping sieve.
> Daddy gone 1 week now."*

So ended Phyllis's entry for the day.
Carla added in her schoolgirl print:

> *"Making new life plans and goals. Concerned about boyfriend,
> Horse and 2 dogs. Can't sleep, mind occupied with thoughts of food.
> Made list of restaurants where food is to our liking."*

There was another list of food labeled "FOOD CONT." Starving as they were, they let their imaginations run wild, listing a huge variety of foods. Once they knew real hunger and their caloric intake was nil, they began documenting more obsessions with dream foods:

"Chung King
Rosa Rita
Diet rite
Root better
Squirt or Quench or grapefruit
Maionaise
Cookies-oreos-buttermilk, Peanut Jumbles
Trisket
Bacon thins
Triangle thins
Wheat thins
Spaghetti sauce and makings
Rice-a-roni chicken
Macaroni and cheese
Tomato soup
Mushroom soup
Mixed salted nuts
Vandyne Chocolate covered nuts (walnuts, cashews, filberts)
Light dipping chocolate
Chocolate covered creams
Chocolate carmels
Blums square mints
Small toffee can
Cameo wafers
Frosted animal crackers
Graham crackers
Round chocolate wafers
Vanilla wafers
Lorna doones
Bugles."

CHAPTER 10

Letters

DAY 14

Friday, March 24, 1967; Diary:

"Beautiful, clear blue, cold, sunny day. No more food or pills, on a strictly ice diet. Carla's feet improving. Phyllis's feet same. Carla is an optimist; Phyllis is a pessimist. Trying to decide who will get valuables if we die. Intend to write letters concerning inheritance. Late in evening plane flew over us, but didn't see us."

The wondering ended. The reality of not being rescued crept into the diary notations by the end of week two. After fourteen days the food had run out. In that space of time they'd lost up to thirty pounds. Few aircraft had flown by low enough to even be considered search craft. Al was long gone and likely had not made it to safety. Waiting another week had confirmed their suspicions and they began to write their last letters. Fortunately they were written before the really bad times came upon them.

One letter, addressed to Phyllis's brother Robert Hausheer in Chicago, reads:

Dear Rob, Elsie and kids,

This is a peculiar letter—if you receive it, I am probably dead. Carla and I are sitting in the wrecked plane on a snowy mountain near Mt. Shasta. We were en route to SF to pick up a Delta flight for Dallas for a week. We crashed Sat. March 11th at noon. Al left the following Thurs. to find help. I think he is dead. Carla and I survive on our fat and snow. No food or heat. Temps. In the 30s. Our injuries are really negligible, aside from frostbitten feet. (I may lose a foot if I am ever found alive.) So we can understand why we haven't been found. We should have perished in the wreck were we intended to die. Carla and I tried taking measurements today with yarn and a ruler and feel we have lost 3-5 inches and 20-30 pounds so far.

Man this is the way to get to know each other! If we are ever found we know lately what our every move will be in the future. We can hardly keep our minds off of food. We find keeping our aching tummies full of snow cuts down on the pain. It's our feet that really have us buffaloed. We don't dare to try to find civilization as Al did—the snow is armpit deep and treacherous. We are weak and would perish I am sure the first night. We have seen only 2 planes in 2 weeks—doesn't that State of Calif. have a search and rescue outfit? Our poor animals at home—The neighbor is watching them for us.

Have been thinking about my valuables, what few I have. Carla is my beneficiary and she's here with me. What about insurances—15000 life for me. 5000 for Carla, jewelry—probably should come your way as my brother. That my diamond ring should go to one of Joe's kids. Al's son, Al jr. should inherit all the financial headaches.

This is awful sitting here starving away wondering how and how soon we will die. We can probably go on another week this way, unless gangrene of toes sets in.

We are writing letters to keep us busy and to fill you in on the details and hope whoever finds this wreck will mail the letters. I am sorry I could be no help with Arch. Haven't contact him for weeks, but will write him one of these letters. We had planned to come to Evanston this summer too. Guess Ellie or Judy or Jennifer will have the sapphire lavalier. The ring is in the safe in our basement. I would almost like to see Steven have the piano-but I guess that will be Ruth's to dispose of.

It's getting dark and cold again so we'd better crawl under our blanket we were fortunate to have with us. Oh for a bath, a gallon of orange juice, a warm level bed, then some solid food. Boy are we going to eat ourselves sick if and when.

My love to you all, you have been good faithful family. I do hope your life turns our better than mine, from here on.

Phyllis

She also wrote a goodbye letter to her former mother-in-law, Mrs. Budd Clarke Corbus in Evanston:

Dear Mom,

Today makes it two weeks Carla and I have been plane wrecked near a mountain top near Mt. Shasta. That means no food or heat, and it's been cold. Carla found a stream but with chest high snow. I don't let her go for water very often. We are living on our stored fat and snow. Al left a week ago Thurs. to find help. Haven't heard from him since so we gather he is dead. Carla and I stayed with the plane hoping to be spotted by a search and rescue plane but no luck. We have had almost continuous snow, rain, fog and freezing and you can't search through that very well. Last evening a plane flew over and he would have seen us had he looked down. I don't think he did.

I am writing this hoping if someone ever does find the plane after we are dead they will mail this to you. Carla suffered a wrenched back and various cuts and bruises. I think I have a broken arm, a bruised leg, bruised face. Now we have frozen hands and feet and I fear mine are gangrenous. Carla's will be OK if she gets help soon. We are replanning our lives and food fills our minds. We wonder how soon we will die. So far we have lost 20-30 pounds already, but are in remarkably good shape.

Have been thinking about my valuables of which I have few. I think one of Joe's children should get Bill's engagement ring. It is in the safe in the basement. This was the one carat diamond set in platinum and rubies. This of course only if Carla dies with me. My crystal and china are negligible, and I wonder who would want my silver. My watch and present diamond are here with me. We have a horse, 2 dogs, a house, a car, a pickup and 2 hotels. What on earth will happen to them? I am writing a similar letter to my neighbor who is looking after the dogs. (We were en route to SF to pick up a Delta airliner for Dallas, Texas for one week) I suppose Al's son, Al jr. will fall heir to the bulk of the estate.

Carla has been a gem thru all this and optimistic too. She feels if we were'nt killed outright, we were meant to live. If we can just hold out til someone spots us.

One last thing—thank you so much for helping us out financially with Carla. There is still some 200 of yours in my checking account. She just took up skiing and liked it a lot. She says she is sorry she never became the LADY you would have liked her to be but we think she's a fine girl—much too young to die.

See you in another world—maybe

Love, Phyllis

Day 15—Saturday, March 25, 1967

By now N9833A had been missing for two weeks and the search planes were leaving the sky, though officially the search effort was still on via the Hamilton Air Base Rescue and Recovery Center. To most people, the possibility of anyone surviving this long in the missing plane was nil. Alvin Oien Jr., however, still believed his family lived, and he pressed onward with search plans. If everyone else was quitting, he'd fly himself. He decided to rent a small 150 horsepower Piper Super Cub and fly it from the Yreka airport. Looking at the map in detail once more, he plotted his grids. He'd comb every inch of Northern California if he had to.

In the wrecked plane, letter writing continued as the searchers threw in the towel. The following was written for Al Jr. and his wife Sally:

Dear Al and Sally,

What a fiasco this trip turned into. By now, Carla and I are dead too. On March 11th, we ran into light snow at 32 degrees between Jones omni station and Red Bluff. Some guy got on the Red Bluff radio and said he got through at 10,000 feet on top of the weather in 5 minutes. Al gambled and lost.

We were at 11,500, not on top and iced up. We came down iced up with no control and bellied into this snowy mountain near Mt. Shasta at 1215 Our altimeter reads 4300 feet and the compass says 6. We are just to the west of the airway and hear commercial planes flying over all day across the other valley. Only two planes passed through our valley in two weeks. There has been lots of snow and rain and it's been between 30 and 40 degrees. Al started out for help the following Thursday to try to find help. We feel he must be dead by now. Our injuries were minimal, but our feet froze now and gangrene is setting in. Have lost 30 pounds by now. No food of course, only snow and hope has kept us going. We are writing letters hoping if this wreck is found and investigated, they may be used as an explanation.

It certainly is hard sitting here waiting to die by freezing, starvation or gangrene. Carla is great. She feels since we survived the crash we were meant to live. We have about $300 dollars cash, the movie camera, plane tickets, my rings and 2 watches, pearl pendant, pearl ring and Al's ring. I don't know about a will but there is insurance and an estate. I suppose by now you've looked into all this. Horses, clock, cars, orientals, hotels, guns, cameras, etc. Sewing machine money. Number to basement safe is 4R, 25, 3L, 45, 2R, 94, 1L, 2. There are 4 keys to inside doors looped together in the middle drawer of the long table under the horse picture in the tin container.

Carla and I are planning a new life, mostly centered around food. We hope to eat ourselves sick. Oh for a bath, hot chocolate, hamburger, level warm bed, pain killers for our feet. Hope our animals can be disposed of satisfactorily. Mrs. Sevenson across the street was supposed to do the honors for the week we were gone. At first if found, we wanted to go to Texas as planned. Now we just want to go home as fast as possible if we are found.

We wonder if California has a search and rescue outfit. If so, they sure did not try hard to find us. All those searches Al flew at the drop of a hat. He even found a missing plane one time. Don't they have to have evidence for insurance? Being abandoned is a hell of a thing. Al was even on a flight plan of all things and always flew the airways. Why are we so difficult to find?

We love you both and wish you a happy continued life.

Love, Phyllis and Carla

The diary entry for the day written by Carla read:

"Overcast but very high—enough room to fly. Hardly any planes— even routine ones. We have hopes of being seen this weekend if the weather stays good. Have started writing letters to the relatives. Feet and stomachs complaining bitterly."

They'd completed two weeks on the mountain.

Easter Sunday, March 26, 1967

Diary, Phyllis:

"Easter Sun. Beautiful day. Absence of planes. We have seen 2 planes in 2 weeks. Doesn't California have a search and rescue outfit? Our snow supply holding up well. We have a mountainfull."

Just when outside conditions changed for the better, their chances of being found by an organized search and rescue effort ended, though their hopes for a rescue remained strong. On the evening of March 26, Hamilton AFRRC officially called off The Viking search, with "no sign of the missing aircraft reported." Toward the end, the searchers' energy just sputtered out, the planes flown less and less since the weekend before, when a large number of planes were flying. The search event proved unsettling for all involved.

From the beginning, the horrible weather conditions made systematically checking the terrain nearly impossible. Precious time had slipped past as the planes waited on the ground. After the storms passed, it seemed no one could have survived in the wilderness during March 1967 in Northern California.

Last-hour efforts to locate the missing Cessna had been flown by CAP Squadron Composite 34 of Eureka, the planes and pilots spending the final two days operating from the home base at Murray Field in Eureka instead of Chico. Twenty-seven of the members and their planes expended another 42 hours looking during the last two days of the official search.

Up north in Medford, SAR Chief Robert Kagy announced the end of the search, explaining another look would take place in late April or early May for the Oien's plane, as well as Art Ward's plane. Kagy did not mention that it was pointless to believe that anyone survived. The melting snow of spring would reveal both aircraft. His plan for a later search implied a recovery, not a rescue.

For two weeks the climate had made the task of finding a lost plane that had navigated along a public airway daunting. Human efforts were not spared. Statistics released to the public showed that over 350 flights were launched looking for the Oiens' plane. Over 550 flight hours were spent covering the giant expanse of mountain country, all conducted in the rapidly changing flight conditions of early spring. The aircraft used varied from the twin engine HU or SA-16s to the small puddle-jumper variety used by the CAP volunteers. Three ground search teams had been dispatched to check on several reports of low and possibly downed aircraft. Hundreds of CAP volunteers participated in the efforts. The record also shows how Al's Oregon Air Guard unit pitched in to help their brother, flying over 65 hours from March 13 until the bitter end on March 26, searching in vain.

In the diary Carla wrote for the 27th:

"Beautiful day and warm. 45 degrees. Heard a plane working to the west of us but didn't quite make it to our hill. Maybe he will make our canyon tomorrow. Spirits high, stomach and feet low."

Al, Jr.'s Lone Vigil

The aircraft engine's sounds and the warmer temperature tantalized Carla and Phyllis into believing there was a search and rescue effort nearby, but the reality was that no plane was looking for them to the west. The only search flying after Easter was done by Al Oien, Jr. in the rented Piper Super Cub, still far to the north. Al stubbornly refused to give up hope. Years later, when asked how much time he spent flying the small Cubs, he just says he flew from morning to night. Stopping only for fuel, he'd return to the sky and resume his lone vigil. His fuel receipts prove the hours. He'd resolved to look until he ran out of time or money.

Al himself laid out a grid on his map from Medford south with dimensions of 30 by 60 miles. Methodically covering each inch of the grid, flying low and slow in the Piper, he covered the almost two thousand square miles himself, in the end flying for over 108 days. He considered the Cub an ideal aircraft for the search. It was light and maneuverable enough to get a good look at the ground, and he explained that he'd found numerous airplane wrecks while searching. Each time he'd encounter wreckage, he'd carefully take the Cub around again and again, checking to ensure what he'd spotted was not the family plane. Sometimes it took three or four passes to verify the wreckage. He discovered the State of California used a different grid system for marking aircraft crashes than the military did. Frequently identification and location became complicated, although he was sure each found wreck was not the one he sought.

Back in The Viking, at least Phyllis the realist knew the likely outcome. In her letter to her brother, she'd mentioned they could go on for another week "like this." Well into that week, the final letter-writing continued.

To the Old Man's brother in Ladner, South Dakota, she wrote the bad news:

Dear Percy,

When you receive this letter, Al, Carla and myself will be dead. We crashed the plane into a mountain of snow on the way to San Francisco on Mar. 11. The following Thursday Al tried to walk out for help—we never heard from him since. The snow is armpit deep and treacherous. Carla and I have stayed with the plane hoping to be discovered, and in 2 and ½ weeks we've had only snow to eat, and seen only two planes. Don't know how we will die—exhaustion, freezing, gangrene of frozen feet. Think it's trying to hail now. Wonder where Al is—he was really very good to us and we had good years.

Carla insists that since we were not killed outright and have only minimal injuries that we were meant to live and will be rescued. I'm beginning to wonder. We can't even cry anymore. We are trying to plan a whole new life and naturally think mostly of food. Oh for a cup of coffee.

Say goodbye to the Nelsons, Plunketts, Haggens and the rest of the natives for us. We always felt welcome by them. I'll write to Hazel too. Haven't anything else to do I guess. Toodle doo –

My Love, Phyllis

DAY 18

Tuesday, March 28, 1967

Diary, Carla entry:

"Snowed this morning but by late afternoon there was sun and blue sky with high winds. We hear our plane drowning around but we can't see him. We must also be hard to see because of a layer of snow."

The goodbye letters continued. To Don and Joan of the West Portland Riding Academy, Carla wrote:

Dear Don and Joan,

When you receive this letter, Mom, Dad and I are dead. We were on our way to San Francisco in our private plane when we crashed in the mountains March 11. We have lived so far on snow for food and drink and Joan you would be please to know so far we have lost over 30 pounds each and have eliminated at least five inches of excess.

I am writing to you because I don't know the Dick's address so please fill Nancy in on the following: Helen had written a check paying all her debts. And gave it to Robin H. Which I guess went through the wash. She should be informed and write another one. In my dresser at home there is a check from Marion Kerns which is made out to me. I suggest she cancel it and write another. Also last month's bank statement in on my dresser. The address of the club should be changed so someone else will receive all information. The name of the club should also be investigated. I thought Nancy changed it to West Portland Junior Hunt but if I remember correctly, the letter from the bank still read West Portland Riding Club.

Please tell the kids I enjoyed holding the office of President of the club and I felt it an honor to be selected to occupy the job. I know that Carrie will do a fine job. I reference to my future plans for club activities Jim will know all of the details because we worked closely together.

In the time I have spent with you I have learned an awful lot. In regards to Terri whoever handles the estate will see to her. Life was nice while it lasted but ended too soon.

Sincerely, Carla

Later she added the note:

PS we must have lost at least 60 lbs. each 6 and ½ weeks with no food.

Carla continued her writing that day, completing a long letter to her boyfriend Rick; the one who spent his last money on a bus trip for Medford, then stood by helplessly as the search geared up and then seemed to finish before it even started.

Dearest Rick,

I don't know how to begin this letter because I don't know if I will be alive or dead when you receive it. If I die I hope whoever finds my body will mail this to you. If I am alive I will mail it myself but will let you know that I am alright. I am going through the most traumatic and ironic circumstances that I have ever been in and it may be my last. No matter what happens darling I will still love you after I am gone. I better start explaining myself before you go out of your mind trying to discover what I am talking about. On our way to San Francisco we were caught in a snow storm and crashed into the Siskyou Mountains. No one was seriously hurt which was an absolute miracle. Dad thought that he had broken ribs on both sides, a broken vertebrae and broken arm. He had three big cuts on his forehead that were gushing blood. Mom cut her leg and ankle. She thinks she has a broken arm and bruised nose. I am the least hurt of all. I might have a broken rib or two and a sprained ankle but it isn't much. I am also having occasional trouble with my neck and sometimes find it difficult to breath because my chest seems to collapse. So much for these details. The airplane is a complete total and there is hardly any room to move. The first night was a fiasco. Mom was delirious until the next morning. Sunday we tore the inside of the plane apart as best we could to try to get some room for us to lie down. By Monday it was still snowing. Naturally we haven't seen planes so I was bundled up and set on my pretty way to find a logging road and get help. Any logging road would have done. The trick was finding one. It was snowing so hard I could not see 20 ft. in front of my nose and I was miserably lost. While walking through snow that was waist high uphill and downhill, I lost both of my shoes and couldn't find them to save myself. After wandering around for 7 hours, in the storm I said to hell with this. If I am going to die I would rather die with mom and dad. The walk with bare feet in the snow did not do my feet any good and I came up with chill blains and frostbite throughout and sores all over my feet so I couldn't touch them.

You sure hit the nail on the head when you said it would be snowing and I would be miserable. All the food we had between the three of us was two packs of sour mint drops and two packs of M and Ms. When we really got desperate we discovered three jars of jelly that mom was taking to Sally, 3 packs of Beemans Gum, three packs of throat lozenges and a few bisodol tablets.

It snowed heavily Tuesday and Wednesday so we didn't go outside the plane. We relied upon the snow for water but that wasn't very appetizing. Our food ration was six lousy M and Ms three times a day. It poured down rain Wednesday night and Dad trapped a lot of the water in the canvas that we covered our entryway out- the right front windshield with. I had been foolish enough to pack my bathing suit and it was a lifesaver. We kept the water in the cap and had our first drink in five days. By this time our M and Ms and fruit drops had expired and we opened a small

4oz jar of plum jam. Since it had stopped snowing I was bundled up again wearing mothers size 10 shoes because my dainty size 7 and a halfs were quite swollen. I was gone just about an hour when I got stuck and the snow froze around me. The snow came up to my chest so I of course panicked. When I finally got myself dug out I went back to the plane. Dad was going out the other way to see if he could find anything. I might add that the mountains were about 8000 feet high and we were at 5000 feet so you see the problem. During the day we saw our only airplane. The other times it has been so cloudy, snowy, rainy or foggy to see more than 50 feet. We are just teased by having to listen to the big planes flying over the top of us on top of the clouds.

We haven't heard anything from dad so we are almost sure that he is dead. The weather doesn't improve but we still keep hoping. The main reason we still hope is that since we didn't die in the crash then we must be going to live. We are having terrible gusts of wind that shake the whole plane. Dad had estimated them to be at least 50 mph when he was here but they are almost twice as strong now. We have maps and what have you shoved in the broken windows and sprung door and baggage compartment. We each have on seven layers of clothing to keep us warm. Without dad it isn't as warm as with only a blanket and a half. We tore one up to make new leggings for me. But there is a little more room. Mom has frozen her feet also and we are worried about gangrene. Out feet are very painful and our nerves are steaming on edge.

Mom and I have had many heart to heart talks and community cries and now have reached an understanding that we never had. She knows our true feeling towards each other and she knows about the ring and our plans for marriage when I am eighteen. She says she doesn't mind but it is my dad who would object. We think dad is dead so we are trying to plan our new lives—if and when we are rescued. If we get out of here you will became a part of our family much the same way I am part of yours. You would be able to come over to our house quite often and even eat with us. I can't wait to get out of here and be with you again and start my new life of understanding and love. You said that you really had a chance to find what life meant to when you were in the car accident. Well it is now Thursday the 23rd and this is the last day we have any food. Mom and I have gone for one week on 4oz of jelly and we are very hungry. I have had two weeks to realize what life, love and you mean to me. I don't want to die but don't have much to say in the matter.

Rick, I love you more than anything else in the world. Even if I die I will die loving you so my love is eternal. I want so very much to be your wife and to raise your children and I hope to be given that privilege. If I do get out of here I don't know if I will get out with all my toes. I hope so. We have lost so much weight my clothes fall off of me. There is still plenty of room to run my hands through my waistline. You have said I look pretty good in a two piece except for my large legs. I've lost all the excess fat on the inside of my legs and from what I can tell they are as slim as the rest of me. I can't wait to see what I look like in a dress or bathing suit, even jeans. It has gotten pretty cold but we can only look at the thermometer when it gets warm enough to go outside or it quits raining, or snowing. This is only during the warmest part of the day. We don't really know how cold it gets during the night. It has gotten as cold as 20 degrees as far as we've seen and that is cold enough. The temperature averages between 30 and 35. But it still is cold. At night we try to snuggle together for warmth. But that doesn't always do the trick. I spend all of my time whenever I am wresting, which is most of the time, thinking of you and if you would pardon the insult—food. Please don't blame me when I only get a teaspoonful of jelly per day so food is on my mind constantly. I can almost say I have not eaten in two weeks because I don't call the toothful of sugar food. When I am back the thing I most want to do out on dates—incidentally I will see you more often—is go over to the snack shack a Barbecue Bakery where Linda works and have a cozy dinner of hamburger and fries. Right now I plague myself with the thought of food. We have made vast lists of places we want to go to dinner. It is now almost 430 and I am almost curled in two because my tummy is so empty. In another half and hour we will finish our jelly and then from there on rely on snow as long as we last or get rescued.

I don't know how to express upon you the depth and sincerity of my love except to become your wife and bear your children. Remember that upsetting topic that I was talking to Linda about when she and Deanna came home from work? Well there is no need for me to be concerned anymore because the worry has been cleansed.

Remember Rick no matter what happens I love you now and

forever more. Dead or alive you will be in my thoughts from now until eternity.

All my love, Carla

PS Please go on to college and don't enlist. It's probably too late to say this now. I don't like this letter but couldn't write another. I changed dates as often as possible.

DAY 19

Wednesday, March 29, 1967
Diary, Phyllis:

"Cold night 20 degrees. High wind blowing high clouds. Only warmed to 35 degrees. Feet less bothersome. Obsessed by chocolate + cheese + ice cream. Carla cold, dizzy, upset—feels period imminent. Have done considerable crying—we just must get out of here."

Phyllis wrote to her father, Walter Hausheer:

Dear Dad,

Here it is 7AM on another dreary day when searching will be practically impossible. If you get this letter Carla and I will be dead. We are waiting to die in the plane wreckage on top of a snowy mountain near Mt. Shasta. We've been here 2 and a ½ weeks living on our fat and snow—no food. Al left the first Thursday trying to make it to help. Apparently he's dead as we've never heard from him again. We've seen two search planes who haven't seen us.

Our injuries were minimal and we are in remarkably good shape except for our frozen feet which are now becoming gangrenous. The snow is armpit deep and treacherous. We felt our best bet was to stay with the plane. At least here we have best bets to be seen and shelter.

We are trying to be optimistic and are planning a whole new future. Carla has been great. She insists that since we were not killed in the crash we were meant to live and be rescued. The temp is 32 degrees and might cold. We've lost at least 30# so far. Would like to see what I look like in a dress. We are writing these letters hoping that is and when anyone who find the wreck and investigates they will mail the letters.

I don't know what else to say but that it looks like I'll exit before you do. It hasn't been an easy life and I hate to sit around and helplessly watch a 16yr old who is full of hopes and aspirations. Well, tally ho. Nuf said.

Our Love, Carla and Phyllis

Keep the lavalier for Judy

CHAPTER 11

Fading Voices

DAY 20—Thursday, March 30, 1967

Now the cold weather returned to torture. The warmed air of the previous day proved to be just a tease. Phyllis wrote:

"Light dry snow ½ the night + morning + afternoon. Completely covered with 1' new snow. Can hardly open hatch. Found vitamins toothpaste + milk of Mag tabs to eat. Brushed hair. Despondent."

The hatch was the canvas engine tarp rigged over the broken right side of the windshield hole. As the snow and ice accumulated on the cover it became difficult to remove. In their weakened condition, just moving the frozen cloth required greater and greater expenditures of limited and depleting energy. Keeping the tarp in place was essential to survival. Sealing the fuselage kept them safe from the raging storms.

Despair crept into the diary as the volume of days approached Week Three on the mountain. It was as if Shoemaker Bally desired to keep the fuselage's location a secret. First, the entire search effort missed their spot, likely due to the mountain's weather. Only two search planes visibly passed by during the first two weeks. Then the incessant snowfall buried everything under several feet of a white, cold, blanket.

If only a streak of clear warmer weather could move over Northern California, they could sweep the wings and maybe set out some distinct markers on the snow, assuming they could even walk anymore. They had become so weak since the crash. Any thought or hope of walking to safety on their own was long gone. But maybe April would bring warmer air and better search conditions, and the return of the cold was just temporary.

DAY 21

Friday, March 31, 1967
  Diary, Carla:

> *"It snowed all night and morning, have over 1½ feet of snow. Afraid we will be completely covered. Started snowing again late afternoon."*

DAY 22

Saturday, April 1
  Diary, Carla:

> *"Have 3' new snow. Dug out our entry. Feet getting better. Blue sky now. Spirit and hope soaring. Our April fool is on fate. Surely will be seen in the next week."*

DAY 23

Sunday, April 2
  Diary:

> *"Beautiful sunny day. No Planes. Worked on meal planning and house redecorating. Aren't they looking for us at all? Health seems to improve surely we're meant to get out of here."*

Using their minds, trying to spend their time productively, the two continued the renovation plans for their Portland home that they'd begun on Week Two. They'd made a list of how the house decorating would proceed, down to the tiniest details:

> *"Yellow paint for the woodwork. Orange or royal blue for nook curtains.*
> *Armstrong floor of blue and beige with coordinating Formica counter*
> *Wallpaper? Or washable paint*
> *Clean Orientals/drapes during painting*
> *Fix vent for 1st bathroom*
> *Clean and wax floors*
> *Keep ceilings white*
> *Do bathroom, kitchen floors & formica first*
> *Carry yellow woodwork color down basement hallway*
> *Floor lamp*

Round table lamp shade
Remove recliner chair and loveseat
Master Bedroom
Blue Oriental rug
Blue pull drapes over panels
– remove Venetian blinds
Some blue quilted spreads
Ecru woodwork
Blue flower wallpaper with ecru background
Carla's 2 beds and dresser plus bedside table & 1 new mattress
Carla's room
Mint green walls and slightly darker woodwork
Dark green rug
Yellow curtains over blinds
Yellow quilted spread
New single bed, table, dresser of light wood
Chair—no desk
Refinish shelves to match or remove it
Or low chest to go under window
Living room and dining room and hall
Paint rosy buff
Solid red rug to go with red in oriental for dining room and hallway?
Yellow bathroom with royal blue towels and shower curtain, etc.
Armstrong tile floor with gold and blue flecks"

They carefully added drawings of each room to their house renovation plan. As they passed the point of having spent one month on the mountain, the entries became shorter. The reality that they'd never leave became quite clear to them, and they expressed the pain of abandonment in their writings.

DAY 24

Monday, April 3
    Diary:

> *"Beautiful sunny day all day—heard 3 comercial planes. I guess no one flies in California."*

DAY 25

Tuesday, April 4, 1967
  Diary:

  *"Sunny in the morning and mixed rain and snow in the
  afternoon—that's just what we need, more snow. No planes."*

DAY 26

Wednesday, April 5
  Diary:

  "Snowed part of night and all day long."

DAY 27

Thursday, April 6
  Diary:

  *"Snowed all night and all day. Can't open the front canvas, the
  snow is too heavy. Our only hope is good hot sun to melt the snow."*

DAY 28

Friday, April 7
  Diary:

  *"Finally stopped snowing sometime during the night. Still can't get
  the canvas off. The weight of the snow is pushing down the cabin
  and it is getting hard to move around. The sky is overcast but the
  altimeter is ½ past. Are hoping for better weather. We have been
  here one month today."*

DAY 29

Saturday, April 8
  Diary:

  *"Got the canvas open finally. Sky clearing somewhat. Occasional
  sun."*

DAY 30

Sunday April 9
    Diary:

"Rain and hail all night and all day"

DAY 31

Monday, April 10
    Diary:

"Rain and hail all night, snowed all day"

DAY 32

Tuesday, April 11
    Diary:

"More snow we think we are completely buried and cannot possibly get out. Our hopes are dubious."

By this time they'd lived in the fuselage for a month straight. For one month they'd not slept in a bed, been warm, or eaten a meal. The amount of calories they'd consumed was miniscule. Their fat stores were gone. Their bodies now ate their own muscles in order to live just a bit longer. Their arms and legs shriveled, and the internal muscles of their hearts and other vital organs wasted away.

As the starving person's body diminishes they interact less and less with the surrounding world. The diary now became a series of brief daily notes as their lives faded. They were done crying, but still they hung on...

DAY 33

Wednesday, April 12
    Diary:

"Still unable to open the tarp. Many planes. Wish they could see us. Roof getting lower."

During the crash, the big, strutless wing cracked through the rear part of the fuselage. As the 2,000-pound aircraft settled further and adjusted with the constantly shifting field of snow, the fuselage turned slightly, twisting the wing and pressing down on the cabin, decreasing the headspace further. Because the floor buckled inward in the crash, the already cramped space, not much more than that of a large refrigerator, became smaller with each shift of the wing. The Viking had become their tomb. Buried underneath snow, there was nothing to do but record the darkness and altimeter settings.

By now the spring winds and sun were melting the snow in the Trinity Mountains, even on the high north slopes, but the aircraft was deeply buried under the snow. While snow melted above them, the water dripped onto them through the broken plane's body.

DAY 34

Thursday, April 13, 1967
    Diary, Carla:

> "Still can't open the canvas—it seems froze solid. Cockpit is sinking but rest relatively the same. We are dripping like a sieve."

DAY 35

Friday, April 14
    Diary:

> "Very quiet day—dark but altimeter going down."

DAY 36

Saturday, April 15
    Diary:

> "DITTO—sun in afternoon."

DAY 37

Sunday, April 16
    Diary:

> "Spit at us in morning—no planes. Dark day."

DAY 38

Monday, April 17
    Diary:

"Almost pitch-black altimeter rose."

Trapped in the dark cockpit now under a deep layer of snow, Carla noted:

DAY 39

Tuesday, April 18
    Diary:

"A little lighter out. Altimeter still up. Don't care if we live or die but wish either one would hurry."

The starving person in a catabolic condition is often too weak to sense thirst and they become easily dehydrated. Movement becomes painful due to the muscle weakness and cracked, dry skin. Atrophy of the stomach causes food cravings to stop because the hunger sensations end; this happens when the section of the stomach which senses hunger shrivels. Vitamin deficiencies which lead to a host of other diseases come into play: anemia, pellagra, and scurvy are common. Lethargy and apathy characterize this level of malnutrition. It was willpower alone that kept the diary active and the two women alive. Each day there was a small entry, an effort to keep interested in life, but by now the entries were brief, like text messages with minimal information.

DAY 40

Wednesday, April 19
    Diary:
    "A little lighter altimeter down slightly"

DAY 41

Thursday, April 20
    Diary:
    "A dark day..."

DAY 42

Friday, April 21, 1967
  Diary:

  "Dark day, getting snow… first altimeter down, then up slightly"

DAY 43

Saturday, April 22
  Diary:

  "Dark day—altimeter same—no grand and glorious change."

DAY 44

Sunday April 23
  Diary:

  "Ditto"

The warmth of spring brought more trouble rather than relief. The tipping plane, now dripping like a sieve, was more and more uncomfortable. The diary closed with sad cries for help.

DAY 45

Monday, April 24
  Diary:

  "More planes than usual—will we be found? Now worried where to get snow. Roof sinking or plane tipping a tiny bit every day. Roof starting to leak more. Please hurry someone!"

DAY 46

Tuesday, April 25
  Diary:

  "Brighter Day—we wanted today to be 'the day' but so far it looks like it is going to be 'another hell of a waste.'"

DAY 47

Wednesday, April 26
Diary:

"Altimeter improving—now at 25 after. We hope to be rescued before or on Sunday—since Sunday is my 16th birthday."

DAY 48

Thursday, April 27
Diary:

"Dark day altimeter ½ past. Wish the rescue [would come soon]"

DAY 49 & 50

Friday-Saturday, April 28-29
Diary:

"Brighter day—sun we think—some passing clouds. Altimeter back down. Tomorrow is my 16th birthday."

DAY 51

Sunday, April 30
Diary:

"Today is my 16th birthday—I wanted to be rescued today—it would have been nice, very bright day, lots of sun. Altimeter as low as it's ever gotten. Maybe we will get out of here soon."

In Portland, Oregon, Rick Katterman thought of his Carla. He had reset his mother's engagement ring with a new stone, and this was the day he planned to present it to Carla. They were then to announce their plans to marry when Carla turned eighteen.

DAY 52

Monday, May 1
Diary:

"Altimeter same. Sunny only. Wish something would happen."

DAY 53

Tuesday, May 2, 1967
  Diary:

  "Bright and drippy day. Altimeter 25 after."

DAY 54

Wednesday, May 3, 1967
  Diary:

  "Very bright drippy day—we are soaking wet. Our beds are puddles."

DAY 55

Thursday, May 4, 1967
  Diary:

  "Bright drippy day—we are completely soaked"

The diary ended on this note, except for days circled on the calendar to May 9. On this last circled day, a final note was added in a shaky, uneven hand:

  "We probably lived longer than the calendar shows but were too weak to mark the days."

After this last entry, one of them summoned a last bit of determination and packaged the Airman's Guide and letters.

The goodbye notes were secured so the first person to discover the wreck would find the stack of papers. Near the top was the note from Carla chiding the search and rescue units saying:

  "Rescue Team. Aren't you proud of yourselves? Look at the days we waited for you—no food, just snow and hope kept us alive the circled days.

  With a flight plan, airports, omni and airbases all around us, how come we didn't see any planes? I think that you deserve a Brownie button for being the most thorough search and rescue team in the country."

At some point near this Carla laid her head on her dad's leather map case and she died. No one knows who passed first, the mother or the daughter.

While Carla Corbus and Phyllis Oien lived their last few days, the Civil Air Patrol sent out notice of an exercise to train over unpopulated areas of the Livermore Valley. Over the weekend, Group 20, comprising the over 300-strong CAP members of the Bay Area, would participate in search and rescue exercises simulating downed aircraft. Twenty-five aircraft would participate in the exercises, coming from throughout California, some coming down from Ukiah. The training would consist of flying aerial search patterns, air-to-ground communication, air drops of medical equipment, and supplies to simulated air crash targets.

Up in Portland at Jackson High School, the tanks of tropical fish Carla had cared for in Mr. Barr's class all were dead. Rick Katterman had stumbled through the rest of his senior year and resigned himself to life without Carla. There would be no marriage to the girl who suited him so well. Stunned, for a summer job he found a position as a fire lookout in the Washington wilderness—withdrawing into solitude. Sometime during the summer he falsely heard the Oiens were found, but by then in his heart he knew the truth.

Al Jr. went back to flying the line for Delta Airlines.

CHAPTER 12

Shoemaker Bally

May 1967

THE HIGH COUNTRY FINALLY came to life again after months of cold, wet winter days. In May, the deep snow on Shoemaker Bally's north side melted, filling Grass Valley Creek to the brim as it bubbled down the steep slope toward Highway 299, eight miles to the north. The scrub brush and chaparral sprang back tall and the brown hills slowly turned green. Birds flew through the high country pines, chirping at nothing as the sun warmed and dried the earth. The occasional black bear and cougar roamed the mountains now, searching for food. Coyotes ran wild up there too, their howls echoing over the long spaces. Voles scurried through the underbrush, rustling dead leaves as they foraged. Deer began browsing the higher ground, weaving their way through the impossibly tangled brush of the frying pan country.

The vista Al Oien, Sr. surveyed with despair before his lonely walk had changed from one of white snowy peaks to friendlier miles of green mountains as far as the eye could see. The blue dome of sky above carried warmer, survivable air, now that the summer sun swung higher in the western sky. If there were clouds in May, they were good weather cumulus clouds, not the icy caps that had obscured the peaks throughout the past months.

The snow visible from the crash site lay mostly on the sides of Mt. Shasta, fifty-five miles to the northeast. The mornings were clear and cool, far different than those tormenting, cold, foggy, windy mornings of the past seven months. The air smelled of pines and brush, clean and fresh, and in the morning, dew covered the plants, only to evaporate with the morning sun and breezes. If The Viking had crashed at this time of year, survival of all would have been guaranteed.

The muddy logging roads opened as the snow melted and their surfaces became firm enough to drive upon. As the days warmed, cars, trucks, and motorcycles used the logging roads again. Citizens trekked out to see the countryside, camping, hiking, and hunting. Commercial airliners still flew high in the sky over the valleys. Occasionally a small plane passed by, paralleling the highway eight miles away, en route to Weaverville or Redding or Red Bluff.

The high mountain winds rocked the old airframe of Alvin F. Oien's once-beautiful airplane. Now The Viking was just a hulk holding the final secrets of the disappearance of a good family. With the snow melted, the problem of camouflage ended. The aircraft lay in plain sight now, the gray wings contrasting sharply with the green hillside, just waiting to be seen from a passing plane.

The Boss did such a good job of bringing her down intact. To those with knowledge of aircraft types, it was recognizable as a Cessna 195, with its elegantly shaped wing and tail combination that lay against the mountain slope. The Viking still looked like a complete plane, something not often said of an airplane that has crash-landed in the mountains. Most end up looking like aluminum foil shot onto the rocks from a cannon. But no planes flew over the wreck. There was not much reason for anyone to fly low over Shoemaker Bally, other than an occasional shortcut or diversion around some weather.

In the hull of the plane, the smell of death lingered. With the melting of the snow, it had tipped far over on its left wing, the one with forty gallons of fuel inside the fuel tank.

The plane lay at her steep angle and sank deeper into the brush as the latticework bushes sprang back to life and stood tall. It was as if the mountain vines tried to hold the plane tighter and tighter, but it did not matter much anymore. The crash survivors were survivors no longer. The wild animals had found the remains and chewed the flesh, scattering bones, clothes, and hair around the crash site.

The Trinity Mountains were known as an active bear area, where the State of California dumped off rogue bears that bothered the human population. Bones and clothing lay intermingled outside the once-proud machine. Clothes, long hanks of dark hair, and piles of bones picked clean were all that was left of the beautiful girl who missed her boyfriend and wanted to be a chemist and who had a horse named Terry, and of her mother, the tough, practical-minded nurse from Chicago.

Out of the Yreka and Montague Airports, a stubborn man refused to throw in the towel. Al Oien, Jr.'s lonely sky quest continued as he wheeled over the mountain peaks and valleys in the rented Piper Super Cub.

Through the rest of the summer when he could, he returned to California and took the rented plane into the sky. Al kept to his own search patterns, burning hundreds of gallons of fuel and meticulously covering each square inch of the countryside. As always, uncountable patches of snowmelt glinted in the sunlight like metal airplane parts. He'd circle overhead, verifying the shiny spot was not a plane wreck, sometimes making pass after pass. He looked for scars on rocks and trees, and anything that looked out of place. Stubborn like his father, the old Air Force Search and Rescue pilot in him just could not give up looking. Junior knew his own father would never throw in the towel easily, and Junior still expected he'd find someone alive.

He'd make sure he fulfilled his charge. Hour by slow hour and mile by grueling mile he completely checked the sectors of his map. He knew what one hundred percent coverage meant and his definition was different than that of the Civil Air Patrol with its two week search. He'd said a few things to the Civil Air Patrol too when he saw them during the search, and had not had an ounce of help from them since March 25. The Viking was out there somewhere in an area that had been "one hundred percent covered." He knew if he stuck with his methods and ideas he'd eventually find it. The Boss was his father, for good or bad, and he'd keep looking for him even if it took a lifetime. Al flew on from morning until night, fueling when needed, not keeping track of anything other than the country he covered.

Overall, California was dotted with more than six hundred air crashes. Many of these were old military wrecks. Frequently there were two documented locations for the same wreck, or the wrecks spotted were ancient accident sites completely forgotten about.

As the snows melted, Junior would find a wreck site, only to discover it was already documented one way or another. Usually SAR parties marked wreck sites with painted yellow crosses to show they'd been documented. In those pre-GPS days, coordinates varied widely.

Because of his search efforts and the time he'd taken from work, he'd been under pressure from Delta Airlines' Chief Pilot, an old hard-to-get-along-with Naval aviator, to give up his futile search and return to the line in Dallas. But Al kept his aerial plodding when his time allowed, working with his brothers to manage the family affairs, and looking for the missing plane. There was the hotel to deal with, the Oien home, and the family's animals.

On May 11, two days after the last day circled on Carla's calendar, Al Jr. stopped searching for a short time. He returned to his normal Delta Airlines schedule in Dallas, but not before having a close call that almost killed him. Through this experience, he found out what likely happened to The Viking.

Through the experience of a close call that almost killed him, he found out what likely happened to The Viking.

While flying from Yreka and searching the rough Klamath River Canyon, Junior's Piper Super Cub was forced higher and higher to keep clear of the mountain peaks. Clouds built around his plane in all quadrants just as they'd done on the day the Viking flew south of Medford. If the plane descended, it could strike a mountain, so his only choice was to fly upwards to the blue sky above the cloud tops. The 150 horsepower Cub struggled to stay on top of the clouds. The son of Alvin F. Oien dodged and weaved his airplane around cloud buildups while heading to the north in an effort to get into safe air.

Suddenly the high clouds engulfed the small plane and Al Jr. was forced to fly using instruments. Unfortunately, the rented plane carried no instrumentation for blind flight.

Al, trained to fly in million-dollar all-weather airliners and Air Force planes, was instantly thrown back into the technology of the 1920s with the Super Cub, and it almost cost him his life. Without the capability to fly into clouds blindly, and struggling for control, Al kept the Cub's wings level, airspeed and altitude constant. Carrying only primitive instruments, the aircraft was unstable and hard to control, having been put in a situation it was not designed to handle. Suddenly the small plane tipped, entering a spiral. Al recovered from the spiral not once, but twice, gambling on which direction to turn.

As the altimeter unwound, he knew he could easily strike the cloud-covered mountains. Fortunately, the Cub popped out of the clouds over a valley near a town with a railroad track passing through. After he became oriented, he recognized it as Yreka. Flying carefully back to the airfield, he set the plane down and parked it. Pondering what had just happened, it struck him that the scenario was likely the same as that encountered by his father's plane.

Unlike his son, the missing senior Oien was not trained as an instrument pilot. His plane had the capability to operate in clouds and carried the proper instruments, but Senior never obtained his instrument rating. He'd always flown it by outside references, using Visual Flight Rules. Instrument flying, known as Instrument Flight Rules, was a different ball game.

After his close call, Al reasoned that in all likelihood his father's plane plowed into the cumulus clouds and flew through icy, Pacific Ocean-soaked air after it passed the Medford Airport on that ill-fated day. Ice formed on the wings and disrupted their airflow, creating extra drag and weight. Just to keep flying and to stay alive his dad was forced to descend toward the mountains hidden by the icy clouds.

Al's nerves were shaken from his close call and he knew he was lucky. The death spiral is a killer. Popping out through an overcast was far better than striking the terrain. Fortunately for him, he'd had enough altitude to clear the terrain and he recovered control over the valley, guessing twice which direction to turn the plane for a proper recovery. Before he flew it again, Al had words with the owner of the Cub and insisted that he install a fifty-dollar instrument called a turn and bank indicator, which would allow a better margin of control while flying around mountains and clouds. The owner installed the instrument. It consisted of a ball in a curved glass tube that allowed a pilot to keep the aircraft coordinated without referring to the ground. Coupled with the airspeed gauge and altimeter, a pilot should be able to stay out of trouble, Al reasoned. He kept using the Cub for searching when he could, and he now believed he knew why The Viking went down.

The Bollings

Northern California and Southern California are two completely different countries socially and geographically. Northern California is rough and mountainous, logging and mining country. On the coast, commercial fishing is the industry that fills wallets. Inland in Shasta County, timber was the cornerstone of the economy in the late 1960s. Movie stars and sandy beaches were few and far between. Loggers and millworkers kept the wheels of commerce rolling. These blue-collar workers came from all over the country and enjoyed a good standard of living. The excellent climate, good jobs, and recreation opportunities made the California–Oregon country a paradise.

A couple named Bolling lived in Shasta, California. Hilda Bolling (née Woolsey) ended up out west in the 1930s as a young girl, arriving on a Greyhound bus from Kentucky along with her ten brothers and sisters. Her father, a soldier in the Great War, finally got his mustering-out pay and used it to make the big move from Kentucky to Oregon. During his war service, her dad had met a buddy from California who'd let him in on the secret of the border country between Oregon and California: The hunting and fishing were phenomenal and jobs were good. Hilda's family sold out, loaded up, and moved on west. With the economy tanking in 1930s Kentucky, things could only get better for the family out on the coast.

In the 1940s, Hilda met Floyd Bolling of Grants Pass; she hated to admit it, but they'd met in a bar one night. Later she married him and ultimately the couple raised two daughters. After serving in the Navy during World War II, Floyd became a debarker, a worker who removed bark from logs at a lumber mill. He and his wife loved the outdoors and lived in Shasta. Together they'd take drives into the Northern California hills during the deer season and hunt together, or sometimes they'd head out to a lake and fish. The two were an experienced, rugged outdoor couple who knew how to enjoy the wild country and put food on their table.

Summer

Spring turned to summer in the Trinity Mountains where the storms of the winter existed only as memories. Instead of cold winds, now, even up high, the air was warm, sometimes even balmy. Instead of deserted country, tourists flocked to the nearby Trinity Alps and Whiskytown Lake for boating, hiking, and camping.

Above the peaks, fair weather cumulus clouds floated against the pale blue sky while the sun burned the hills into dangerous brown baked dryness as June turned into July, then August. Sometimes thunderstorms and showers formed in the later afternoons. The once-soaked terrain became endangered by wildfire, and parts of Northern California burned terribly each summer as lightning struck the dry timber.

The hulk of The Viking lay in plain view now, still undiscovered. People traversed the road up to Bully Choop Peak, passing the wreck by just a few hundred yards. Being the highest peak, from its summit a person could look down upon the entire scene of the two-month drama. Oddly, no one saw the plane.

When he could, Al Jr. continued to fly his search pattern out of Yreka, grinding away from morning to night, maximizing his time aloft. He'd refuel and take off again without a break. At night he'd fall into bed at his motel, exhausted from handling the Cub at low altitude for hours at a stretch. Between search flights he'd return to Dallas and Delta Airlines to fly the line again. Sometimes Ron came down to California too, to check on his brother's search progress.

Al Jr.'s flight grids eventually moved south, and in August he passed just five miles north of The Viking, the closest he ever came to his family the entire time they were missing. In all, by himself in the rented plane, he searched a total of one hundred eight days. He could not believe the family, especially his father, was dead. He needed proof.

Life moved forward. The Clifford Hotel was under new management with the Old Man gone. Chuck's emergency leave from First National Bank was not renewed and he was forced to leave his banking job. He ran the Clifford for three months straight, and all through the summer, except for the two weeks of his Oregon National Guard duty. Ron was doing double duty, working at Sears and helping run the hotel. They had found an attorney to help handle their dad's business affairs, with no known will in existence.

While the mystery of the Viking remained open, another was solved. The remains of Art Ward's CallAir plane were found on a mountain up near Ashland, Oregon on July 21 by some hikers. Ward's CallAir was the missing plane from December, which the

California CAP sought so hard a few months before the Oien search. In the newspapers, Ward's crash was attributed to weather. "At the time the aircraft became missing a storm was reported to be moving through the area," read accounts of that crash.

CHAPTER 13

Found

October 1, 1967

Iɴ ᴛʜᴇ ꜰᴀʟʟ, Cᴀʟɪꜰᴏʀɴɪᴀ deer hunters still drive along the Bully Choop/County Line Road searching for black-tailed deer. Getting up high along ridge tops allows hunters to look down at the country, giving them the high ground advantage. Ancient Bully Choop Road has been an ideal place to scan for deer because of its elevation in relation to the surrounding land. Though it's frying pan country, the deer run near the roads too, and there are brush-free zones on the hillside, giving hunters clear shots and easier access to their game. Sometimes a lucky shot brings one down in a clear area, eliminating a lot of work when packing meat out.

On a crisp weekend day in early October, Hilda Bolling and her husband, forty-two-year-old Floyd Bolling, left Shasta for Bully Choop Peak in their Rambler. The Bollings knew deer hunting was good around Paradise Hill; they figured on taking a look and spending the day driving along the ridge tops of the Trinity Peaks. Though she did not have a driver's license, Hilda drove slowly along while Floyd glassed the slopes. The morning passed as they enjoyed the mountains, admiring the view and searching for animals.

At mid-day, after stopping their station wagon up on the steep, narrow run to Bully Choop Lookout, Floyd looked to the east with his high-powered binoculars. Below the summit of Shoemaker Bally, in a brush-covered depression, he noted a shiny spot that looked like metal. At first he thought it could be an old wrecked airplane.

He said to his wife, "Hey, you know that kind of looks like an airplane over there!"

Hilda scoffed and said it was "more like an old sun-bleached log!" when she took her turn glassing the slope. "Let's take a look anyways," Floyd said.

They came back down the road, closer to the spot, and Floyd set out across the ravine with his rifle while Hilda waited in the Rambler. He knew he was in bear country and did not want to be unarmed. Fighting his way through the chaparral and buck brush, he fell three times and broke his rifle stock. The going was so difficult he almost turned around several times. Working up a sweat and panting, he finally came close to the gray object he'd seen from above. He did not spot it until he was about twenty feet away. It was a gray and blue plane with a rose painted on the tail, sunk deep into the brush, left wing tipping downhill with the nose pointed east toward Redding. Floyd Bolling had just found what hundreds of people had not: The Viking.

At first glance the Old Man's plane looked intact, "not the way a plane should look which landed in the mountains," Floyd later said. But as Bolling fought his way through the brush to look closer, he saw the fuselage was indeed badly damaged. The one-piece wing was broken on the back of the plane and the left wing lay downhill at a steep angle. The right wing was crumpled. The windshield was smashed on the right side. The belly had caved in upward, crunching the fuselage. A prop blade was slightly bent. The plane had struck hard. As he approached, Bolling smelled death. Whoever was in the plane had been dead for a long time.

He shouted loudly for Hilda on the road below to come up and see what he'd discovered. Hilda worked her way to the site, leaving her rifle behind, and joined Floyd at the wreck. She too smelled the odor. Both could see it was not a recent accident. Outside the fuselage they found piles of dark hair and human bones. Some bones were covered in women's clothes and two ladies' purses laid outside of the plane. The smell of decomposition overpowered them. Floyd knew they'd need to call the sheriff and thought they'd best get some identification from the plane. He grabbed a suitcase and his wife took a small plastic box from on top of the instrument panel, neatly wrapped as if put there for someone to find right away. The smell almost knocked them over and they hurried to get away from the odor.

The couple fought their way back to the station wagon, stumbling downhill through the chaparral that had confounded Carla's feet, while carrying the box and suitcase. Even without an impossible layer of snow, transiting the mountainside was difficult for healthy, well-fed people. Finally reaching the station wagon, Floyd drove back down the hill fast toward Redding.

Along the old narrow road they ran into another hunter they recalled as MacPherson, who had a radiotelephone in his truck. He offered to let Floyd make the call to the sheriff.

The radiotelephone couldn't reach Redding due to the high hills, so MacPherson and Bolling moved to higher ground for better reception and finally got through to the Shasta County sheriff. After notifying the authorities, the Bollings drove back to the site to wait for them, giving the couple time to look through their stunning discoveries.

In the late fall sun, seven months after The Viking had disappeared, the Trinity Mountains were a different world, a place of peaceful beauty. It was there and then Hilda and Floyd Bolling discovered the final sad chapters of the Oien family. Gasping from shock, they read from the papers Hilda had pulled from wreckage.

"Whoever finds this wreck—please mail these letters for us. We waited so long for you—where were you? Our daily log is here for you to see in the folded Airman's Guide."

There was the note on pink stationery with two owls up in the left hand corner.

"SEARCH AND RESCUE
Aren't you proud of yourselves? Look at the days we waited for you—no food, Just snow and hope kept us alive the circled days."

All of March was listed in calendar form and circled, starting on the 11th. Below the calendar was the continued note that said:

"With a flight plan, airports, omni and airbases all around us, how come we didn't see any planes? I think that you deserve a Brownie button for being the most thorough search and rescue team in the country."

On the right side of the page, the calendar for April listed all the days through the 30th, with the 30th circled with a starburst. Next to the burst it said:

"This would have been my 16th Birthday (over)"

On the backside was a shakily written May calendar with the days circled to the 9th and the numerical list written to the 13th.

Below the May calendar, Carla's note about being *"too weak to mark the days"*—that is, the days lived following May 9th.

Below the cover letter were lists of foods and restaurants, then the letters, stamped, addressed, and sealed.

Hilda and Floyd Bolling sat in the fall sun on the bench seat of the Rambler, waiting for the sheriff while reading the words written in the Airman's Guide in red-colored ink about the long deaths of the mother and daughter.

Hilda described the reading, "as if they were right there, speaking to you." She and Floyd together counted the days on their fingers, and as they did, the horror of it all struck them both. For eight long weeks Phyllis and Carla struggled to stay alive up on the mountain, eating snow, just hoping to be found. Somewhere out in the wild country, Al Oien, Sr. set on his impossible way to find help through deep snow. Who knew where he could be now, the Bollings wondered as they surveyed the valleys and peaks all around them. They considered the terrible fate of the mother and daughter: eight weeks unseen and forgotten, marooned in the snow on the side of Shoemaker Bally.

> Hilda described the reading, "as if they were right there, speaking to you."

After meeting the sheriff and turning over the diary and suitcase, the Bollings returned to Shasta and never again drove along the ridge tops of the Trinity Mountains. The grisly find bothered Hilda for a long time. Over the years she saved every news item she found about the crash. Sometimes she'd sit in the kitchen of her modular home in Lookout, California, and re-read the clipping about the Oiens' missing plane and how she'd helped solve the mystery.

Response

The media chased the story. John Zamboni was ready for anything. Twenty-year-old Zamboni, a photographer and engraver for the Redding *Searchlight*, took a call from a friend that Saturday afternoon. The friend had heard over the county sheriff's radio that an airplane crash was just found out near Buckhorn Summit, west of Redding. Zamboni tore his way into Redding, grabbed his Nikon cameras and headed west to Buckhorn Summit in an ambulance with his friend. He met the gathering officials and the Bollings on

Bully Choop Road. Deciding to team up for the rest of the venture, Zamboni drove to Shoemaker Bally with a Shasta County deputy.

Bushwhacking his way into the impossible brush with the group of officers, Zamboni related how "You could not see a thing until you were twenty feet from the plane!" The group of men found the plane and just stood there for a moment, gasping at the find. Then Zamboni set up his cameras and went to work. He took the first photos of the crashed Viking. Later he described the bones, hair, purses, and found out a skull lay about eight feet from the plane. Deputies pulled out the luggage and noted it was heavy with the odor of decay, indicating the two ladies most likely passed away inside the cabin.

Remnants of brown hair showed Carla Corbus probably died with her head on her dad's leather flight map case, using it as a pillow. Deputy Jerry McCarthy read from the diary, and the officers collected the evidence and later turned it over to the Redding sheriff.

Immediately following the discovery, the brothers were alerted and Ron Oien jumped into his brother Chuck's car and sped down to Redding from Portland. Of course, Al Jr. flew there and together they went to the crash site with the coroner, George Files. Alvin Oien, Sr.'s wallet, containing the five hundred dollars listed in one of Phyllis's letters, never was recovered from the accident scene, a painful memory that the Oien boys couldn't forget. It was not the money but the photos Al Sr. carried that were irreplaceable.

Initially the reports indicated three bodies were recovered, but once investigators inventoried the bones and read the diary, it was clear the Old Man was still missing.

The next few days were a blur of activity. The Oien brothers, along with thirteen other men—trustees from the Shasta County Jail or deputies—searched down from the crash site. Initially the plan called for searching two miles down in a rectangular pattern from the site. Though searching in bright sunlight, they returned totally wet after beating the brush, which was soaked by rain. The needle in a haystack search, as reporters called it, needed more men. "Hope is attached to the deer hunters who were swarming through the country," the reporters wrote.

Later in the week, Weaverville Sheriff Tom Kelly put two hundred and forty Job Corps workers on the search. They lined up five feet abreast and walked north through the valley all the way to Highway 299, eight miles in all, without finding a trace of the Old Man.

Funeral and Obits

Carla and Phyllis were buried together at Forest Lawn Cemetery in Portland during a private ceremony on October 10. Their obituaries were separate in the columns of the Oregonian. With different last names, a person had to read the announcements to realize Carla and Phyllis were mother and daughter.

> Oien—Phyllis Jean of 8523 SW 56th Ave; wife of Alvin F. Oien Sr., Portland; mother of the late Carla Jean Corbus; daughter of Walter E. Hausheer, Willamette, IL.; sister of Robert Hasheer, Willamette; Memorial Services will be held Saturday, 10 a.m. Chapel of AJ Rose and Son, Eastside Funeral Directors, SE 6th and Alder.

A phone number was given and "Private internment Lincoln Memorial Park" followed a "no flowers please" comment at the bottom. Carla's obituary listed the same time and place as Phyllis's, with her survivors as the Oien brothers, her grandmother, Mrs. Budd Clarke Corbus, and the Hausheers in Illinois.

Phyllis and Carla had finally made it home, but the whereabouts of Alvin F. Oien, Sr. remained a puzzle that his sons vowed to solve. Though it had been such a long time since the crash, they still felt the Old Man would come walking out of the woods.

CHAPTER 14

Aftermath

NATIONAL TRANSPORTATION SAFETY BOARD (NTSB) crash investigators pulled the Jacobs radial engine from The Viking and sent it to test facilities. The tests showed the engine was making power when the plane crashed. To keep onlookers from intruding on the sanctity of the crash site and to avoid repetitive reports of a plane wreck, the hulk of the Old Man's pride and joy was skidded down the mountainside by loggers and buried in a pit along the roadway. The sons considered salvaging the plane, but ultimately decided against it. The Viking had been damaged beyond repair. Junior did recover the instrument panel (including the glove box), and Chuck saved the door emblazoned with its name: *The Viking.*

Speaking to reporters about the ruggedness of the crash site, Al said that when he carried the instrument panel down the mountain, he "fell out" rather than hiked.

During the final cleanup of the crash site, maps, the empty jelly glasses, and a coat with the odd label, "Wear In Good Health" (an old Irish blessing) were recovered. Some of Carla's bones were found under the fuselage. Somewhere out in the brush on the steep slope, never found, were the shoes Carla lost on her first excursion when her feet got frostbitten. Years later, Ron Oien expressed displeasure with the quality of the cleanup. He felt that the amount of the found bones indicated a disinterest on the part of the authorities, that they could have done a better job of it just out of respect for the victims.

While in Redding the Oien brothers declared to reporters they'd spend the rest of their lives, if needed, combing the Trinity Mountains and the surrounding area. They'd do whatever it took to find the Old Man and bring him home. They considered him alive until proven otherwise.

Junior's experience as a professional search and rescue pilot with high-powered and well-equipped all-weather aircraft greatly influenced his outlook on the Civil Air Patrol. He publicly criticized the CAP as a group who were only "well-meaning amateurs who could not put out a maximum effort search," and who, if they were properly organized, would have found his father's aircraft. For one thing, the search was commanded out of Chico, fifty miles too far south, when it should have been closer to the Trinity Mountains where the plane was found, not far from the airways the pilot reported he'd be flying along. The Associated Press sent Al Jr.'s comments throughout the world, eliciting a response from the CAP.

Major Hillier of the USAF WARRC in Hamilton is quoted as saying of the search efforts: "It was a little bit more than we normally put out in such bad weather." He rebutted Al Jr., saying the mission was "well planned and flown to the best of the CAP's ability." He added, "We feel bad there were people out there that we left. There were bad snowstorms and it was cold. No possibility of life existed in the conditions. It was hard to find a gray aircraft against snow. We are horrified to think someone lived." There it was in black and white. "No possibility of life existed." Obviously the searchers never knew the resiliency of the Old Man and his family.

CAP officials spun out varying statistics of their involvement in the search. Major Hazel Smith, another CAP official was quoted: "Over the two week search period in May, the CAP flew 415 sorties with 45 planes, involving 165 people." Smith claimed the winter and spring had been one of the wettest and coldest in California history, making the task of finding the gray plane very difficult.

Carla's story became worldwide news. Press agencies, magazines, and newspapers clamored to get a copy of it. Initially the diary contents were disclosed through a Portland attorney to family members only. Then the Hausheers and Oiens hired an agent to deal with the publicity sparked by the diary. Eventually, partial contents of it were published in the January 1968 *Saturday Evening Post* article.

Journalists from around the world flocked to the crash site, some even camping there, claiming to be able to see and hear traffic on Highway 299 and to see the lights of logging trucks driving the hills, implying that the Oien family just sat on the mountain and watched traffic pass nearby. They paid no mind to the fact no logging occurred in the winter in that area of the Trinities, or that the unmaintained roads became impassable to wheeled vehicles once the snow piled on the high country. Not once did the diary mention wheeled traffic.

Floyd Bolling was interviewed by the *Oregonian* and commented that the survivors had been too passive.

"They had three suitcases full of clothes. They could have laid them out in a signal. There was gasoline, matches. The brush would not have broken easily. Probably the women could not have broken it at all. And fire building in the snow and cold could not have been easy. But they could have doused the clothes in gasoline and oil, started fires by igniting the clothes."

He noted they had flares too, but failed to explain they were hardwired inside the aircraft, designed to light off when a pilot trips an electrical switch. Not to mention the flares were imbedded on the left side of the fuselage, the side that was buried in foot after foot of snow.

He failed to consider what Carla and Phyllis would wear once their clothes were burned, or who exactly they'd signal with the fires when only two search planes overflew their site in the first two weeks, back when they still had food and some energy.

In the same article, Calvin P. Burt, a Lake Oswego mathematics teacher, was quoted as saying, "Only recently did I realize I was a survival expert and did not know it." In the past he'd given classes to Air Force personnel on wilderness survival. Burt explained that all people who fly in light aircraft should have survival training. He outlined several ways The Viking survivors could have helped themselves by eating tree bark and small animals. Their task was to make themselves known. They should have made a fire and Al was "asking for trouble" when he set out cross-country without snowshoes. The article concluded with Bolling's statement:

"It didn't seem like they'd done anything at all. They just laid in the plane."

Questions were raised in other news articles about why Al flew through the storm when it was predicted, and why three people with extensive outdoor experience failed to survive such an event. The best answers seemed to come from the Old Man's eldest son, who told reporters:

"There are a lot of good questions. But the answers are all guesses. Why didn't they clean snow off the wing? Why didn't they make a toboggan out of airplane parts? Why didn't they lay out some sort of signal? It's easy for us, warm and dry and in no pain, to look at a map and wonder why three cold, hurt, starving people, lost on a mountainside in the snow didn't do certain things. If they were strangers to me I'd say they were dumber than hell. But I know my dad. He is not dumb. He is tough and calm and level headed."

No one addressed the fact that both adults suffered head injuries in the crash, and that untreated head trauma is not a pathway to clear thinking.

In the media deluge, there was one other person who spoke up for the Old Man: his old friend Walt Rupert, owner of Rupert's Flying Service. The veteran airman seemed to be the only one outside of family who stood up for The Boss, saying he was a good pilot who took good care of his plane and that the plane and engine were kept in good condition. Rupert added he really liked Al Oien, and that he'd seen him take off on March 11 for the last time.

In his days Rupert had seen many pilots come and go. Old pilots are careful to not pass judgment on other pilots. They know that fate really is the hunter.

The Oien brothers launched energetically into their new quest to find the Boss. With the discovery of the aircraft a big question mark hung over everyone's life. Out there somewhere in the mountains maybe the Boss still needed their help. The diary gave them clues as to where he might be found.

In November, with the weather staying clear, Al Jr. took another leave from Delta. In Portland, he and his younger brother Ron loaded up a Volkswagen Beetle with camping gear and a border collie named Tippy they'd just got from the local pound for $50 and drove to Shoemaker Bally. Each day they'd beat the bushes looking for their father. They set up a camp in a sandy spot along the old road, just down from the crash site.

After learning the terrain, the sons formulated a theory on how their dad's walk from the plane proceeded. With a broken right arm, he must have grabbed at branches with his good left arm, causing him to swing toward the west. To the west, the road ran toward a low spot between Bully Choop Peak and Shoemaker Bally, which opened up to more hospitable terrain to the south. Besides, the road below the crash site ran downhill and to the west as well. To get through this low spot required traversing some of the roughest terrain in the west. A person lost up in the mountains likely pursued the downhill option. Unfortunately the direction to the highway along that road was east and uphill.

The boys worked through this area, beating the brush, searching for clues. At night they'd return to the camp, soaked, beat up, branch-whipped, and exhausted. They'd eat, then sleep in their tents, get up, and repeat the search. Their little fifty-dollar pound dog fended off the black bears.

As the cold season returned, they got just a small sample of what Carla, Phyllis, and the Old Man experienced. The temperatures dropped with the shorter days. The fog, the wind, the rains, and snows that marked the signature of winter returned. They knew it was different for them than for the crash victims. They were well fed, in good health, and could pick up and leave when they liked, but they did get a taste of the mountains in winter.

Right after the wreck was found, they'd enlisted the help of inmates from Redding and Weaverville, and teams of men combed the mountains all the way north down to Highway 299, not finding much except for some old logger's clothes and paperwork. Sheriff's Deputy Francis Keeler even exercised the county bloodhounds on the mountain, hoping to catch a whiff of Al Sr. In the end though, the others gave up hope, just like during the aerial search.

Soon, Weaverville Sheriff Tom Kelly refused to allow any more men on the hillside, unless some sort of break occurred in the case. The terrain was too rugged and rough to let people walk through without better leads. "Someone could break a leg or fall and get killed," he told Al Jr. Sheriff Kelly became involved in the case once the crash site was found to be in his county. Initially, response teams came from Shasta County, believing the site was in their domain. The Bully Choop Road was also known as County Line Road. The crash essentially occurred right on the line.

Al and his brothers kept searching on their own into the cold season. Then the story petered out and only occasionally surfaced in the news. The next March, near the anniversary of the accident, Al Jr. even tried to access the site by snow machine, but that proved impossible too. To the reporters they mused about dropping into the mountains by parachute, as a way to get up there under the same conditions as during the plane crash. Frying pan country just closed down in the winter, locked in ice until summer.

CAP Evaluated

Less than ten days after the Bollings found The Viking, the Civil Air Patrol was ordered to Fresno for its annual evaluation. Three hundred members were ordered to participate in a simulated search for a missing aircraft. The simulated test under "actual" conditions was to start on Friday, October 20.

The California Wing of the Civil Air Patrol was downgraded in its score from the year before during its annual readiness test in Fresno. The downgrade "had nothing to do with CAP criticism" by the Oien son, according to newspapers. Scoring ninety-one out of one hundred possible points, the main CAP deficiency was noted as "not planning enough emergency fields for the search planes." The evaluating USAF officials complimented the Wing on its "over-all effectiveness." The evaluation consisted of a practice search and rescue scenario conducted under the supervision of the Air Force and directed by CAP Lt. Claude C. Morgan of Burbank, California.

CHAPTER 15

ELT Beacons Work

FOLLOWING THE CRASHES' DISCOVERY, reporters interviewed Al Oien, Jr. on the subject of botched search and rescues. He stated bluntly, "I'd like to see someone get off their dead bottom and develop a locator beacon. There is no real reason a locator beacon could not be developed that will sell for less than $100." He added the beacon would be "like galoshes, you don't really need them but you take them along."

At the time, bulky emergency beacons were used in the Vietnam War by military pilots. The downed flier carried the beacon in a vest and personally turned it on, allowing search and rescue aircraft to home in on his location. This application of technology was not lost to Representative Richard Ottinger of New York, or to Senator Pete Dominick of Colorado, who'd been pushing for some years for emergency beacons to be mandatory equipment for civilian aircraft in America. If private aircraft carried locator beacons, the "search" could be eliminated from search and rescue. By the late 1960s, a few states had mandated ELTs aboard private aircraft.

At the same time, electronics manufacturers were developing just such a device, some with the idea that the emergency aviation frequency of 121.5 would be monitored by the thousands of high-flying commercial aircraft that crossed the United States all day and night. The fact that over 60 airliners flew within a short distance of the downed Oien plane each day of their abandonment was not lost on those who knew of the event.

The following April, when Carla would have turned seventeen had she survived, an article appeared in several newspapers discussing the Northrop Corporation's development of a portable distress beacon which could be used to locate downed planes. The story also mentioned that because of the Oien tragedy, the FAA considered

a

c

making locator beacons mandatory gear for small aircraft in the United States.

The Northrop beacon was a cigar box sized device that weighed seven pounds and could send a signal up to one hundred fifty miles away. A Northrop spokesman mentioned that over 2,000 airliners overflew the Oien crash site during the eight weeks that the family members survived in the wreckage. The Northrop beacon would have pinpointed their location to the high-flying airliners.

Early Locator Testing

A year and a month after the Bollings found The Viking and her sad contents, on a November day with high overcast clouds, a small transmitter called a DAL, for "downed aircraft locator," was placed on the north side of Shoemaker Bally, not far from the spot where Carla, Phyllis, and the Old Man suffered. The device was installed in the tail section of a light aircraft, the empennage being the part of a plane that usually survived a crash intact. The transmitter location was selected at the 6,000-foot level in a difficult-to-access crevice, just to ensure that the testing would be a serious challenge for the overflying aircraft.

a small transmitter called a DAL, for "downed aircraft locator," was placed on the north side of Shoemaker Bally, not far from the spot where Carla, Phyllis, and the Old Man suffered

The search pilots were briefed on a scenario wherein the emergency signal they sought emanated from a private aircraft that disappeared on a flight from Medford, Oregon to Sacramento, California.

The Garrett Company of Los Angeles had faith in this new device and planned to prove its efficacy that morning. As the engineers stood on the side of Shoemaker Bally, two FAA DC-3s departed Red Bluff Airport, equipped with radio receiving gear tuned to the emergency frequency. The lumbering DC-3s flew off to the northwest, where the airborne radio gear picked up the emergency signals sent out by the DAL.

Climbing above the cloud layers to avoid the jagged Trinity peaks, and following the information on the needles of the special radio receivers aboard the planes, the FAA pilots locked on the emergency signal from 9,000 feet. Within forty minutes, the DAL location was pinpointed on a map to within two miles, though the area lay underneath a thick overcast. A second test by the FAA planes pinpointed the simulated crash site even faster.

The following day, five CAP aircraft outfitted with receivers were launched to conduct the same test from Red Bluff. The CAP pilots flew their small aircraft at 10,000 feet. All five pilots located the simulated wreck in an average time of one hour and forty minutes. One aircraft dove through a hole in the overcast and actually spotted the radio-equipped tail section and the Garrett engineers.

In a similar demonstration at the Grand Canyon, CAP pilots located a simulated downed aircraft at the bottom of the canyon within one hour and fifteen minutes. The CAP spokesperson stated the wreckage would likely have been given up as lost due to it being impossible to see from the air location. Distress beacons worked.

The Old Man Found

The years rolled by after the sad discovery of Carla Corbus and Phyllis Oien's bodies in the wrecked Viking. Somewhere, Al Sr. lay below those DC-3 airplanes testing the new emergency beacons. As time passed, other aircraft became search objects for the California Wing of the CAP and the search for the Viking became a memory listed only as "search WARRC # 32," referencing the unsuccessful search for a missing civilian Cessna 195 flying from Medford, Oregon.

Al Jr. flew the line for Delta, piloting the company's DC-8s to the West Coast from Dallas, while Chuck and Ron managed the day-to-day business affairs of The Boss. Still seeking answers and closure, from 1967 to 1969 the Old Man's sons searched the Trinity Mountains for him when they had time to spare—in all, they estimate they made about half a dozen trips to the mountain. They'd drive down to California in a Volkswagen and comb the rough slopes of Shoemaker Bally, protected from bears by their little yapping dog. Without finding proof of his bones, it was impossible to believe their tough "Viking" of a dad was gone for good.

Eventually their belief in the Old Man's survival faded, but the boys insisted on looking until they found him or ran out of life themselves, after all, he was "our dad," as Al Jr. told the constantly prying news reporters.

On September 23, 1969, two-and-a-half years past the day when
The Viking made her last takeoff from the Beaverton Airport, two
Redding men decided to hunt black-tailed deer in the frying pan
country. Hunting carefully along Bully Choop Road that morning,
James White and Hohn Graham discovered some bones and clothes
next to a big log about a half-mile downhill from The Viking's
crash site.

What drew the two hunters' attention that fall morning was the
chance sighting of a glove placed on the top branches of a nearby
tree. The glove stood about eight feet above the ground and pointed
straight up the mountainside to where Carla and Phyllis had died.

The glove…pointed straight up the mountainside to where Carla and Phyllis had died.

So, The Old Man had thought of his wife
and daughter to the end, hoping his glove
would point the way to the plane.

On the ground below the tree with the
pointing glove, along with men's clothing
and some scattered bones, were a plastic
pen and wallet. On the pen it said,
"Compliments of the Clifford Hotel 527 East Morrison St.
Portland, Oregon, 97214, The Central Eastside Hotel, Alvin F.
Oien, Owner."

The rectangular brown plastic wallet was almost chewed into an
oval from small animals biting on the edges. It too read "Clifford
Hotel" on the sides. The wallet held papers that were still intact,
though badly soaked through. Amazingly, they survived two years
of exposure to rain, snow, and small animals. The hunters read the
paperwork, at least what was still legible.

On the receipt for fuel from Rupert Flying Service, Beaverton,
Oregon (Receipt #4451, in the amount of $19 dollars and change),
in blurry blue ballpoint pen, it read something like this:

Last Will and Testament Al F. Oien
My Estate
Please send help
Kennedy
Murcheson
(Accountants and Lawyers)
Bury me next to my daughter at Lincoln.

And there was a mention of the AJ Rose Funeral Home in Portland.

So Alvin F. Oien, Sr. died hard and alone. He'd pushed himself to the end the way he pushed himself throughout his life. Though he'd not made it more than a half-mile from his beloved Viking and family, he died the way a man should die: *trying*.

With broken bones, no food, and only his willpower, he plowed his way through the armpit-deep snow as far as his battered old body would take him. He had found a road and tried to get out, but just could not go any further. In spite of his severe injuries, he'd dragged himself across the county line from Trinity into Shasta County before the end came.

Death may have visited that first night after he left the shelter of the plane, when the fog settled in along with the cold, or he may have lasted longer and expired trying to get back to his wife, daughter, and plane. In the end, he remembered his six-year-old daughter, the one who had succumbed to the childhood disease and he wanted to rest with her. Maybe he believed Phyllis and Carla would make it off the mountain, alive.

Like a final insult to the family that had suffered so much, the Old Man's body was found within spitting distance of the campsite where his own sons based their ground search for his remains. If he'd not placed a glove as a marker in his last act with his last strength, it is doubtful whether he'd ever been found.

The two hunters, Graham and White, drove into Redding and reported their discovery to the Shasta County Sheriff's Office. At 1240 Deputy Ramos, Sgt. Merrifield and Coroner Gatos drove to the site of the bones, located 8.6 long miles from Highway 299, along Bully Choop Road. Searching the area with the two hunters, the sweep revealed scattered bones, clothes, and Al's personal items. His pens, keys, coins, cufflinks, camera, and glasses were all listed in the sheriff's report and turned over to McDonald's Mortuary in Redding. They even found his Optimist Club pin.

Deputy Ramos dutifully completed his report that evening, then turned the case over to the coroner for positive identification, though he was certain the body was that of Alvin F. Oien, Sr., whom he listed as the victim at the top of the report, labeled FOUND AIRCRAFT III-30. In a carefully drawn sketch, he outlined the position of the bones and belongings, showing how Oien may have taken shelter next to a large broken log located only forty-five feet from the road.

In Redding, the Old Man's remains were positively identified through dental records and he was personally flown home to Portland by Al Jr. in a borrowed Cessna 210, and then placed with his family at Forest Lawn Cemetery.

With the finding of the father, open-ended legal matters closed. Later in the month, the Multnomah County courts up in Portland resolved Al's estate. The life work of a man who moved west as a young boy with only a shoebox full of possessions and a desire to work almost fifty years before was valued at $144,000. When taxes, fees, and debts were settled, the three brothers received $81,000 to be divided among themselves.

ELT Bill Enacted

Al Jr. had always insisted that the search effort to find his family was a complete failure. Yet significant events followed their terrible deaths. Senator Pete Dominick, a Republican from Colorado and a 5,000-hour pilot himself, was outraged when he read about the failed search in the *Saturday Evening Post*; "Please Hurry, Someone," as told by Harold Martin, outlined the terrible story of the death diary. The Oiens' story spoke to Dominick about the unnecessary suffering and despair that occurred on the slopes of Shoemaker Bally, while technology existed to find lost aircraft.

Dominick read portions of the diary into the Senate records while he successfully promoted a bill mandating that ELTs be installed in all U.S. aircraft. President Nixon signed the bill into law in 1970.

As a pilot he was aware of emergency radios and their potential to save lives. Adding to the list of air accidents he had been bringing to the record starting in 1967, Dominick read portions of the diary into the Senate records while he successfully promoted a bill mandating that ELTs be installed in all U.S. aircraft. He and Senator Warren Magnuson of Washington were instrumental in pushing the bill ticated of weapons and defense measures. However, that same sophistication is no deterrent to the Communist guerrilla who, as Mao Tse-tung has said, operates as a "fish swimming in the sea of the enemy."

The battle to neutralize the danger of nuclear attack can be won. The battle for men's minds, and world security, is not won. Until it is, the need for nuclear weapons and deterrents will remain.

FEDERAL BAR ASSOCIATION SUPPORTS DIRECT ELECTION OF THE PRESIDENT

Mr. BAYH. Mr. President, I am pleased to invite the Senate's attention to the action taken by a distinguished organization on the question of electoral reform. At its convention held recently in San Francisco, the Federal Bar Association adopted an eight-point resolution favoring an amendment to the Constitution which would provide for the election of the President and Vice President by a direct, nationwide, popular vote. It is gratifying to note that the stand taken by the Federal Bar Association conforms closely with the features of the Joint Resolution the the

observance of National 4-H Week. Nearly 3 million members of 4-H in every State and territory, residing in both urban and rural environs, are now undertaking diverse new projects for the coming year.

As a former member of 4-H, I know firsthand the incalculable contributions being made by th ing head, heart, would, I think, be state the tremen Club activities ha five decades to ru

I think it is als particularly durin the increasing co programs are ma sters. The House mittee, in its rep 1968 agricultural served this fact tional funds and providing adequa relate to person development prob dies.

NATIONAL   AERONA SPACE ADMINIST PRIATIONS, 1968

Mr. MANSFIELD. M unanimous consent before the Senate ness.

The PRESIDENT

CRASH-LOCATOR BEACONS

Mr. DOMINICK. Mr. President, an article in the Washington Post of 3 days ago was undoubtedly noted by many of my distinguished colleagues with a great deal of remorse. I refer to the report of the diary written by 16-year-old Carla Corbus and her mother, whose remains were found almost 6 months after their plane went down in the mountains of California. I only hope the top echelon of the Federal Aviation Administration, who have consistently refused to require installation of crash-locator beacons in private aircraft, will take note of this tragic incident.

At this point, Mr. President, I would like to recapitulate on some of the points I have brought out in previous statements urging requirement of crash-loca-

Excerpt of *The Congressional Record* showing the moment Senator Pete Dominick introduced the Carla Corbus diary into the Senate record. (See the whole record pertaining to Senator Dominick's efforts and how many times he referred to the Oien family tragedy in excerpts from 1967–1970 starting on Page 197.)

through Congress; then in 1970 President Nixon signed the bill into law as Public Law 91-596, as a rider to the OSHA bill.

It was Carla and Phyllis's diary that broke the political gridlock and spurred the government to take action. Since 1970, human lives saved by ELT radios worldwide number in the thousands, perhaps even tens of thousands. With the development of the newer 406 MHz beacons, thousands more will be found. The ELT story is directly traceable to the suffering of Carla, Phyllis, and Alvin F. Oien, Sr. in The Viking.

Years later, when I was visiting Al Jr. that first time and he insisted the entire search mission for his family was a miserable failure, I handed him the following news clip from May of 1971 and told him there were lots more stories like it:

Emergency Radio Beacon Was A Pretty Good Deal

RED BLUFF, Calif. (AP) — Buying an emergency radio beacon that guided Air Force rescue craft to his crashed plane "was the best deal I ever got in my life," says an 18-year-old Kent, Wash., pilot. Jim Myers and three other injured University of Washington students were lifted by helicopter Saturday from the crash site 20 miles southwest of Red Bluff.

The Western Rescue Center at Hamilton Air Force Base said the rescue was the first achieved in its area with the aid of the emergency distress beeper in a civilian plane.

In 1967 a crashed plane in the Trinity Mountains northwest of Red Bluff never was found by air searchers. A diary found by deer hunters with the remains of a Portland, Ore., couple and their 16-year-old daughter revealed they had survived two months before dying of starvation.

The victims in that crash were Mr. and Mrs. Al Oien and Carla Corbus, who kept the diary.

Myers, who crashed Friday night after being caught in suddenly thickening weather, said, "We had no business getting out of this alive, none of us."

The distress beeper device, which Myers bought four months ago at a $144 discount price, was triggered by the crash impact and a search plane picked up its signal Saturday morning. The beeper guided a C130 plane and an H53 Jolly Green Giant helicopter from Hamilton to the still cloud-covered crash site.

A team of paramedics provided emergency treatment for Myers and his companions until the helicopter lifted them to the Red Bluff airport Saturday.

I sat quietly as my friend, who flew alone for over one hundred days in the Super Cub, searching for his lost family, finished reading the news story. Then the old search and rescue pilot just shook his head, as the eyes that searched for his lost family for thousands of hours teared up and he said:

"That's how it should have been, they should have been found… that's how it should have been!"

Epilogue

THOSE COLLEGE BOYS WHO were saved in the first ELT rescue in Red Bluff lived on to become pilots for the major airlines.

The Oien brothers sold the house in Southwest Portland. The trim little home testifies to how we all pass through this life like a vapor. The plum trees of the survival jam still live. The current owners know nothing about how the vibrant young girl and her mother redecorated that house in their minds as a way to cope through the dismal days on a distant peak.

They don't know how, in the kitchen, Phyllis kept the copper bottomed pots so shiny they gleamed like mirrors, or how the Old Man slapped his wife's bottom and announced how much he loved her and belted out his di-dee-do song in that deep baritone voice.

Hilda Bolling was still alive in 2011 and I visited her (she passed away in 2014). The Bollings kept working hard, enjoying the Northern California outdoor lifestyle. Floyd retired and moved with his wife to the remote privacy of Lookout, way east of Mount Shasta. Hilda told me he had passed on a few years back. Then in her eighties, she still dyed her hair and looked like she could still bushwhack through the frying pan country. She had never got her license to drive, said she never learned, though she drove the Rambler just fine up old Bully Choop Road in 1967. To that day, as she told me, she still thought about the airplane crash they found—she still shook her head saying she should have gotten there sooner.

Rick moved on too, after taking the summer fire tower job in Washington state; he had resigned himself to a new life without Carla. Unfortunately he passed away in 2013 but his newspaper obituary reveals he lived his life as a man to admire, his compassion no doubt tempered from his loss.

The CAP still functions as it did during this search, finding lost people and saving lives. None of the records of the search for the

Oien family were available, most likely having been lost to time. The search and rescue operations I wrote about were put together through a conversation with Jim Bigelow and through examinations of the newspaper articles about the crash. The only official record available came from the USAF through a Freedom of Information (FOI) request showing the Hamilton-based Grummans only flew for a brief period before the duties were turned over to the CAP.

In the late '80s the Oien boys sold the Clifford Hotel, after running it for nearly twenty years. The downtown Portland area declined through that decade, then with the re-gentrification of Portland in the '90s, rebounded as a dynamic, vibrant part of the city's scene. For a time the Clifford was used as an apartment complex, then it got completely refurbished and reopened again in January 2011 as subsidized housing for lower income Portland residents.

The Old Man might not have approved of subsidized housing, but no doubt he'd like the way his old hotel looks today. After refurbishment, it looks as if it will be around for another hundred years. The man who carried the financial records in his hat and punched the union organizer is long forgotten. His old office with the funny signs was gutted during the remodel. Writing this story, I got to where I kind of knew the Old Man and loved him. He may have been hell to live with but he did pretty well for a kid who grew up alone in the crucible of a logging camp. Like his son Chuck said: "The Old Man did the best he could. He never had a childhood so he never could relate to his boys." We really are a mix of good and bad... How could you not like a man who flew over needy kids at their summer camp, 'bombing' them with candy?

Ron and Chuck still live in the state of Oregon, enjoying their retirements and families. The sons are philosophical about their tough old dad. Chuck says the Old Man must have done something right as none of the boys "ended up in jail or on a slab in the morgue." They've all lived productive lives and served the country.

Shoemaker Bally stands tall in the Trinity range as it has for millions of years, right next to Bully Choop Peak. In Weaverville, no one recalls the demise of the Oiens, and neither Trinity nor Shasta County possesses official records of the crash anymore. On the mountain's north slope, trees with the branches that the Old Man used to swing his way downhill with his one usable arm are the same trees from 1960s, but they're much taller now.

My wife Kate and I went up there in the fall of 2010. By triangulation from old photos, I found the approximate crash location. Finding the exact spot really did not matter because the slope was all the same: Impossibly thick brush and steep terrain made travel nearly impossible for a man in good shape on a clear and warm fall day. Being so warm and well-fed, I could only imagine a tiny portion of the isolation felt by the marooned family in that long-ago ordeal. The high-flying airliners still plied the airways above, just as in 1967; still so out of reach except by a radio.

I sat holding one of the playing cards cut from The Viking's upholstery, the one Al Jr. gave me on my first visit. I thought about all that occurred nearby, and back to how I had envisioned it when my dad told me of this story so long ago. I reflected on the two beautiful and ordinary American women who once held the very card I had in my hand and how they'd been thrown into such a horrible predicament. I noticed the spot I'd chosen was marked with bear scat. This was bear country, just another element of terror in the long, fear-filled saga of Carla and Phyllis. I'd forgotten that the State of California released rogue bears in the Trinity country.

Grass Valley Creek, where Carla hobbled to collect water, bubbled nearby. I could hear it. The government rehabilitated the creek and rebuilt the road to prevent erosion during the early part of the millennium. The remains of the plane are buried somewhere along the roadbed of Bully Choop Road and I looked hard for The Viking but it's gone, returned to dust. The instrument panel with small glove box door, the logbooks, and the cabin door with the words "The Viking" are all that remain of Cessna N9833A, the real Viking's pride and joy.

A chilly feeling ran through my body up on Bully Choop. I've spent some time as a police officer and had the same feeling at grisly crime scenes. I could not leave fast enough. It was eight long miles back to the main highway, easy in a rental car, an impossible trek by the injured in the winter. No one could've walked far from The Viking in winter, I'm convinced.

Al Jr. continued to fly the line and became a Delta Airlines captain. As if the man had not suffered enough, in 1971 his wife Sally was killed in a car accident in Texas as she drove to her hospital job in Dallas. A drinking teenager in an out of control pickup truck completely destroyed her Volkswagen as she drove to work.

Al later married Carol, who is still a flight attendant for Delta. Carol is a gem of a woman, a southern belle who loves people, and when you visit them, she treats you as royalty. I can only imagine how well she treats her Delta passengers. As Carol approaches well over forty years on the job, she still loves flying for Delta. Al says he doesn't understand why she's still working.

Al retired from Delta at age 50, collecting on a five-dollar bet he'd made when he was a co-pilot. He insisted he flew only because it was a good way to make a living and he'd get out of flying when he could. He maintains that he never was a real airline pilot but he still loves airplanes—he is a first class raconteur of aviation lore, being a member of a diminishing breed of pilots, men who transitioned from propeller-driven planes into the early commercial jet era.

He and Carol now spend their time between Washington and Texas. Over this book project, I've become good friends with Al Oien, Jr. and have respect for the man. Letting a stranger into your life and discussing emotional details is hard to do, but Al and Carol brought me into their lives like a son and the details of this writing would've been impossible without his help.

It is hard to say why this story gripped me, now that I am fortunate enough to have finally solved the mystery—at least up to a point. Maybe I saw the old Post article as a child, or saw something in a newspaper that remained in my subconscious, but I can now say that for whatever reason, the visions I carried in my head uncannily matched the facts and details of the Carla Corbus tragedy.

There are still questions that remain unanswerable. What unseen forces led to this book? How could it be that Al's house lay under the final approach to the Port Angeles Airport, where I first heard of his lost family? There are also the parallels and similarities of our fathers, the pilots: both self-made men and hard-charging individuals. Then there are the questions of the two men's final flights. Why was the Vikings' survival kit left behind by an experienced pilot? And why did my father, a trained physician who knew about hypoxia, leave his oxygen equipment in his hangar? Both men were tough, smart and gifted yet leave behind some wonderment.

Maybe what Al Jr. told the reporters is the best answer to these riddles. "There are a lot of questions and they are all good." No one really knows what went through the mind of Al Sr. as he punched onwards past Medford, Oregon that fateful day, or how a few days later he struck out down the mountain on an impossible mission with his serious injuries. Likewise, what happened with my father? Did he have a glimmer of anything as his aircraft began its gradual twenty-five mile descent into the ground?

But there is one thing I do know for sure and the record shows: Without the Carla Corbus story, Senator Pete Dominick never would have spoken on the Senate floor about the need for ELT regulations and it was the straw that broke the political indifference to that need. Without the shedding of the Oien family's blood, the technology and laws which led to the ELTs that have saved lives, would not have been passed or implemented.

ELTs have been far from perfect, broadcasting far more false alarms than true notifications; nevertheless, the number of lives saved from these relatively inexpensive beacons number in the thousands. Last I checked, the estimate was that survivors had been rescued from over 5,000 crashes in the United States alone, since the ELTs became standard equipment.

Plenty now walk the earth because of the suffering of Al, Phyllis, and Carla, as well as the Oien sons. With the changes in recent years to the 406 MHz beacons, the accuracy of the signals has drastically increased to where the beacons can provide the latitude, longitude, and aircraft registration number. Searchers now can simply call a registered owner when a beacon signal is detected to verify the condition of the aircraft. Satellite searches coupled with GPS technology now ensure air crashes are pinpointed much faster than in the 1970s.

[Title 14 of the Code of Federal Regulations, Chapter 1, Part 91]

§91.207 Emergency locator transmitters.

(a) Except as provided in paragraphs (e) and (f) of this section, no person may operate a U.S. registered civil airplane unless—

(1) There is attached to the airplane an approved automatic type emergency locator transmitter that is in operable condition for the following operations, except that after June 21, 1995, an emergency locator transmitter that meets the requirements of TSO-C91 may not be used for new installations:

(i) Those operations governed by the supplemental air carrier and commercial operator rules of parts 121 and 125;

. . . . .

Some of the FAA regulations that resulted from Congress' efforts started by Senator Dominick, who was moved by the reports of the Oien family tragedy (see the excerpts on Pages 223–225).

There are other promising technologies that will surpass radio locator beacons as a means to find the missing; even so, ELTs help take the search out of search and rescue.

As for myself, I overcame my youthful ailments. Now I do the thing that always tugged at my heart: I fly planes. I've been all over Alaska in small planes and am now a Dehavilland Dash 8 captain for a regional airline in Alaska. Flying over the beauty of this great state every day is about as good as flying gets. I'm grateful, humbled and privileged. The kid on the mechanic's creeper dreams came true far beyond what he could have imagined and if I could sing, I'd belt out the di-dee-do song for you.

A few pieces remained missing, though. For a long time I thought of my father and of his old Cessna 195. After his accident, my mother was forced to sell that beloved plane, but she kept a home for us five kids, a place of stability through an uncertain time. I ended up purchasing that same airplane with my younger brother, David, in 2012. No longer pristine, it needs to be rebuilt. It is the same plane I'd been polishing that day when I heard the strange story of the lost family that I just shared with you.

When I sit in the cockpit, I'm stunned. I swear I can still smell those Swisher cigars. It really was not so long ago that I polished the thing, dreaming of the day I'd be the owner. I got what I sought and to be honest I'm not sure I want the airplane now. It is the old memories of flying and polishing her with my dad, and flying in it with my family, that I cherish—not the plane itself. It's just another thing, an inert hunk of metal.

Like the victims of The Viking, it seems like my father who once seemed so larger than life just like the Old Man, now is an ancient memory too. Mom was recently in the hospital where he practiced and only one old doctor remembered him. "A good guy" is how he was recalled. That's not a bad epitaph and I hope for the same. It took me a long time to forgive him for the pain unintentionally inflicted on my family through pilot error.

I do understand why my father wanted to set world flying records. I understand too why Al Oien, Sr. pushed onward during that final flight when he could have turned around at Medford....

The drive to see things through is innate and ingrained. As pilots and as people we have to proceed with caution, but without risk nothing gets done. Risking it is the human story. The first people who left home put it all on the line, wanting to know what lay out there beyond the horizon. These people never turned back to the comfort of the places or people they knew. They faced danger. Some lived, some died. Some survived to spread the seeds, which became us. Regardless of their outcomes and fates, the knowledge gained is bequeathed to us all.

> *Ex malo bonum. (From bad comes good.)*
> —*St. Augustine, Sermon LXI*

APPENDIX

NTSB Report

Years later, after all was said and done, Al Sr.'s report ended up like this:

NTSB Identification: OAK68A0032
14 CFR Part 91 General Aviation
Event occurred Saturday, March 11, 1967 in SHASTA, CA
Aircraft: CESSNA 195, registration: N9388A

FILE	DATE	LOCATION	AIRCRAFT DATA	INJURIES F S M/N	FLIGHT PURPOSE	PILOT DATA
2-1187	67/3/11 TIME – 1215	NR.SHASTA, CALIF	CESSNA 195 N9388A DAMAGE– DESTROYED	CR- 1 0 0 PX- 2 0 0 OT- 0 0 0	NONCOMMERCIAL PLEASURE/ PERSONAL TRANSP	PRIVATE, AGE 59, 5000 TOTAL HOURS, UNK/NR IN TYPE, NOT INSTRUMENT RATED.

TYPE OF ACCIDENT
   ENGINE FAILURE OR MALFUNCTION
   COLLIDED WITH: TREES

PHASE OF OPERATION
   IN FLIGHT: NORMAL CRUISE
   LANDING: FINAL APPROACH

PROBABLE CAUSE(S)
   PILOT IN COMMAND – IMPROPER OPERATION OF POWERPLANT & POWERPLANT CONTROLS
   MISCELLANEOUS ACTS, CONDITIONS – ANTI-ICING/DEICING EQUIPMENT – IMPROPER OPERATION OF/OR FAILED TO USE
   MISCELLANEOUS ACTS, CONDITIONS – ICE–CARBURETOR
   WEATHER – SNOW
   WEATHER – ICING CONDITIONS–INCLUDES SLEET, FREEZING RAIN, ETC.
   WEATHER – CONDITIONS CONDUCIVE TO CARB./INDUCTION SYSTEM ICING
   PILOT IN COMMAND – ATTEMPTED OPERATION BEYOND EXPERIENCE/ABILITY LEVEL
   PILOT IN COMMAND – CONTINUED VFR FLIGHT INTO ADVERSE WEATHER CONDITIONS
   TERRAIN – HIGH OBSTRUCTIONS
COMPLETE POWER LOSS – COMPLETE ENGINE FAILURE/FLAMEOUT -1 ENGINE
WEATHER BRIEFING – NO RECORD OF BRIEFING RECEIVED
WEATHER FORECAST – UNKNOWN/NOT REPORTED
MISSING AIRCRAFT – LATER RECOVERED
EMERGENCY CIRCUMSTANCES – FORCED LANDING OFF AIRPORT ON LAND

SKY CONDITION
   UNKNOWN/NOT REPORTED

CEILING AT ACCIDENT SITE
   3000

VISIBILITY AT ACCIDENT SITE
   5 OR OVER (UNLIMITED)

PRECIPITATION AT ACCIDENT SITE
   SNOW

171

OBSTRUCTIONS TO VISION AT ACCIDENT SITE
   BLOWING SNOW

TEMPERATURE -F
   34

WIND DIRECTION – DEGREES
   150

WIND VELOCITY – KNOTS
   10

TYPE OF WEATHER CONDITIONS
   VFR

TYPE OF FLIGHT PLAN
   VFR

REMARKS – RECOVERY DATE – 10/1/67. NOT INST RATED

41 ARRSq History, January - March 1967

at Monterey. WARRC Closed the mission when the HU-16 returned
to Hamilton AFB.

(6) WARRC 18 Assist - 9 Feb 67. On 9 February a C-135
inbound to Travis AFB from Yokota AB, Japan, lost an engine and
requested an intercept. A 41st ARRS HU-16 on airborne alert was
diverted to make a maximum-rescue-coverage intercept. The C-135
encountered no further difficulty and landed safely at Travis.
WARRC closed the mission when the HU-16 returned to Hamilton AFB.

(7) WARRC 32 - 12 Mar 67. A civilian Cessna 195 was
reported overdue on a flight from Medford, Oregon, to San Francisco
on 11 March. The alert HU-16 was scrambled for a route search but
returned after two hours due to darkness and maintenance problems.
Other rescue forces and CAP aircraft continued the search for the
next eight days. On 20 March the 41st ARRS rejoined the search and
sent an HU-16 to search an area west of Mt Shasta in northern
California. On 22 March WARRC suspended the mission with negative
results and no new leads.

(8) WARRC 35 - 20 Mar 67. On 20 March a Navy A4D was
reported overdue on a flight from Yuma, Arizona, to Palmdale,
California. A 41st HC-130 was diverted from a training mission
to begin a route search. However, a ramp check at Palmdale
located the missing A4D and the HC-130 was recalled before it
started the search.

(9) WARRC 36 - 20 Mar 67. A civilian Cessna 150 crashed
10 miles northwest of the Napa VORTAC on 20 March. The alert
crew was scrambled with pararescuemen aboard but could not get to
the crash site due to weather. A Coast Guard helicopter in the
area assumed on-scene-commander duties and relieved the HU-16.
WARRC closed the mission on 21 March after the survivors had been
rescued by a ground party.

c. SAR Mission Summary

Mission/Location	Sorties/Hours	Objectives
WARRC 02 - 2 Jan	* -------	Civilian
N. of La Paz, Mexico	+ 5/26:40	Cessna 337
WARRC 04 - 6 Jan	* ------	Classified
N. of Las Vegas, Nevada	+ 2/8:10	

4

A copy of the only surviving "official" search-and-rescue mission report for
The Viking: the official SAR summary record "WARRC #32" from Hamilton
AFB, recorded by the Western Aerospace Rescue and Recovery Center
(obtained by the author via the FOIA). *Continued on next page…*

41 ARRSq History, January - March 1967

WARRC 09 Assist - 25 Jan N.E. Hamilton AFB	*--------- +1/2:00	T-33
WARRC 12 Assist - 30 Jan E. San Francisco	*--------- +1/2:50	Civilian Piper Apache
WARRC 14 Assist - 3 Feb Pacific Ocean	*--------- +1/4:15	Civ Twin Beech Intercept
WARRC 18 Assist - 9 Feb Pacific Ocean	*--------- +1/3:40	MAC C-135 Intercept
WARRC 32 - 12 Mar Medford, Ore to San Francisco	*--------- +2/7:45	Civilian Cessna 195
WARRC 35 - 20 Mar Yuma, Ariz to Palmdale, Calif	*--------- +---------	Navy A4D (False Mission)
WARRC 36 - 20 Mar N.W. of Napa, Calif	*--------- +1/1:10	Civilian Cessna 150

(* = HC-130, + = HU-16) - (There were no losses of ARRS men or equipment).

    d. Training Summary (Upgrading actions completed this quarter)

Position	HC-130	HU-16
I/RCC		1
RCN	1	1
I/RCFE	2	
RCFE		1
FE/RCRO	2	
I/RCRO	1	1
RCRO		4
RCLM	1	
FE/PJ	1	1
I/PJ		1
PJ	1	3

8. Personnel:

    a. Key Personnel:

       Commander - Lt Colonel Cortez C. Brown
       Operations Officer - Lt Colonel Leslie E. Gamble
       Asst Operations Officer - Lt Colonel Charles B. Knudson
       Standardization Officer - Major George E Glaeser
       Staff Navigator - Major Richard L Ladendecker
       Safety Officer - Capt Harry V Boyd
       Maintenance Officer - CWO-4 Marvin T Higginson
       Administrative Services Officer - Capt David J Frazier
       First Sergeant - MSgt Clyde N. Howey

           5

A Family Album:
Photographs and Memorabilia

Carla Corbus, 1966.

Alvin and Phyllis with Carla at their wedding, August 1958, First Methodist Chapel, Seattle, Washington.

At left, top: Al Oien, Sr. with Carla and friend, 1958, with The Viking, N8388A;

bottom: Phyllis with The Viking, 1958…She flew with Al Sr. all over the U.S.

The middle snapshot is a close-up of "The Rose" of the City of Portland on the tail…with an unknown couple, friends of Al Sr.—he just loved to fly The Viking and would take anyone, anywhere.

A logging camp near Raymond, Washington (circa 1920) where Al Sr. started his career as a cook's helper. He worked his way up to choker setter before leaving the woods. He is the dark-haired boy fourth from the right.

Buckeye, North Dakota, mid-1960s: It was a big event whenever the Old Man flew to the ranch to visit his mother and brother Percy; the town newspaper mentioned his visit.

Early 1960s—Carla and her mother loved to ride horses. Carla's bedroom wall was covered with her ribbons won at the Portland Riding Club.

Southwest Portland, Oregon—the home redecorated in the minds of Phyllis and Carla while they waited for help.

Al's Clifford Hotel in downtown Portland…where he once punched a union organizer.

"The Boss," Alvin F. Oien, Sr., hard at work in his office…his beloved airplane in a photo on the wall.

1967: The Oien boys, thrust into the management of the Boss's affairs...Chuck (seated), Ron, and Al Jr.

Al Jr. at his father's desk at The Clifford... wondering... (late 1967).

Trinity Mountains, 1967: Armed for bear, Al Oien, Jr. ponders the fate of his father.

The author now...and back then, with his dad Rod Nixon, and siblings.

Carol Oien, the author, and Alvin Oien, Jr.

Shoemaker Bally Peak, when the author and his wife traveled to the crash site.

Closer-up view of Shoemaker Bally, looking straight towards the crash site, and the small fire road near where The Viking crash-landed.

The "impossible brush" country of the Trinity Mountains. "Frying pan country," the locals called it.

The road near the spot where they found the Old Man's remains.

```
        FOUND  AIRCRAFT  III-30                    69-5743

2.  VICTIMS & WITNESSES: Additional
    V#1 OIEN, ALVIN F., (SR.)  WMA 57 yrs.,8523 S.W. 56th St. Portland, Oregon
    W#5 WHITE, JAMES R.,  WMA 30 yrs., 2145 Sacramento Dr., Redding, 241-7744
    W#6 GRAHAM, HOHN M.,  WMA  63 yrs., 2145 Sacramento DR., Redding, 241-7744
    W#7 MERRIFIELD, TREVE, Sgt. ID BUREAU
    W#8 RAMOS, FRANK #149  SCSO  Reporting officer

4.  INVESTIGATION:
    9-23-69 Tuesday 12:40PM - WHITE and GRAHAM came to the Sheriff's Office
    and reported that at 11:15AM they had found a skeleton of a person while
    deer hunting on Bully Choop Road approximately 8 miles South of 299 W.
    They had a ball point pen that was found by the remains.  On the pen was
    written "COMPLIMENTS OF CLIFFORD HOTEL, 527 S.E. MORRISON ST., PORTLAND,
    OREGON, 97214, THE CENTRAL EASTSIDE HOTEL, 282-7189, AL. F. OIEN, OWNER."

    MERRIFIELD, RAMOS, and GATOS along with WHITE and GRAHAM proceeded to the
    area.  Upon arrival at the scene, it was found to be 8.6 miles South of
    299 W on Bully Choop Road.

    2:50PM - At the scene the remains were found on the West side of a log
    45 feet from the South edge of the roadway.  The remains consisted of the
    jaw, backbone, hip bone, and parts of the arms and legs.  Also a shirt
    and sweater was found.  Approximately 20 feet South of this a shoe was
    found.  In searching the area the skull was found approximately 25 feet
    West of the log in some brush.  About 15 feet Northwest of theskull a
    pair of trousers was found.  A glove was found 8 feet above the ground stuck
    on the end of a branch.

    Reporting officer along with the above witnesses searched the entire
    area for the rest of the remains. Scattered throughout the area was
    found other parts of theskeleton, however, the complete skeleton could not
    be located.  It is unknown at this time what parts of the remains are
    missing.

    Coroner GATOS in checking the trousers found a silver dollar with keys,
    $3.55 in coin, and other property as listed below.

    The remains were put in a plastic bag along with the property and
    transported to McDonald's Chapel.

5.  PHYSICAL CONDITION:
    One human skeleton transported to McDonald's Chapel.

7.  EVIDENCE:
    One ball point pen with inscription of "Compliments of Clifford HOtel, 527
    S.E. Morriston St., Portland, Ore. 97214, The central eastside hotel,
    232-7189, Al F. Oien, Owner"
    One cuff link, brought to the office by WHITE and GRAHAM, retained by
    MERRIFIELD.
    One sketch attached to case.

8.  PROPERTY:
    Found in the trousers and shirt, $3.55 in coin consisting of quarters and dimes
    one pair of glasses, one camera, 1 silver dollar on key ring with keys, re-
    tained by coroner  GATOS.

10. DISPOSITION:
    Case turned over to Coroner for positive identification of OIEN.
                              Ramos, F. # 49
                              9-23-69 Tues. 6:45pm/ph
```

The report that Deputy Ramos filed when he investigated the remains found along the road to Bully Choop Peak.

The Old Man's wallet with edges nibbled by animals, yet found near his remains with some of his last instructions written on a receipt. His final shelter was a large log just a few feet from the road.

Two sides of what was a little playing card, found at the crash site...Phyllis and Carla cut out their "deck" from the airplane seat Naugahyde. (Actual size approximately 1 x 1.5 inches.)

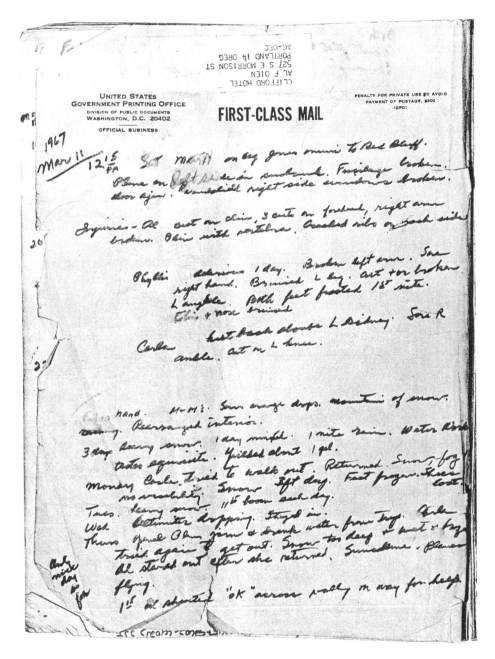

The first page of diary notes were started on the rear cover of the 1964 Airman's Guide.

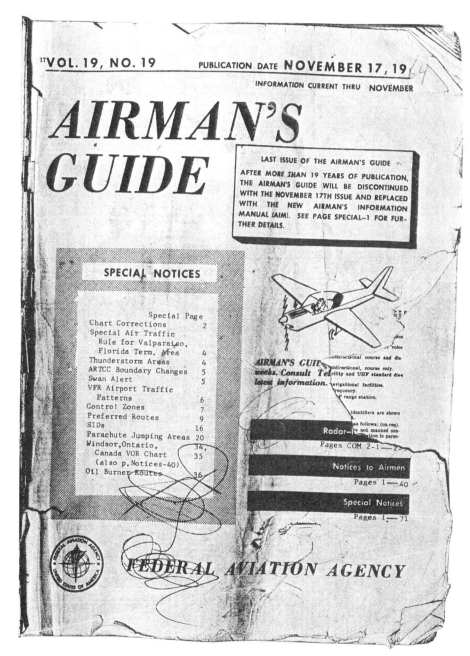

The front title page of the Airman's Guide that served as their diary.

22
Wed   cold, windy, foggy, over cast. Heard a plane last night that could have seen us, we hope. 1 day jelly left. Water gone. Snow now dirty. Feet worsening. Lost considerable weight.

23 Thurs.   Wretched nite. Poured again. Caught some water. Having stomach cramps + refused aches. Feet terribly swollen, toe purple. Reading "Fate is a Hunter." Snow, rain, hail, sun alternately all day — very strong winds. Heard 2 planes today but did not see. The 11:25 explosion was much closer. Cabin is a dripping sieve. Daddy gone 1 wk now.
Making New life plans + goals. Concerned about Boyfriend, Horse and 2 dogs. Can't sleep, mind occupied with thoughts of food. Made list of Restaurants where food is to our liking.

24 Fri.   Beautiful, clear blue, cold, sunny day. No more food or pills, on a strictly Ice Diet. Carla's feet improving. Phyllis's feet same. Carla is an optimist; Phyllis is a pessimist. Trying to decide who will get valuables if we die. Intend to write letters concerning inheritance. Late in evening plane flew over us but Didn't see us.

25 Sat   Over cast but very high - enough room to fly. Hardly any planes - even routine ones. We have hopes of being seen this weekend if the weather stays good. Have started writing letters to the relatives. Feet and stomachs complaining bitterly.

Both women took turns writing in the diary. The difference in their styles is evident…Phyllis generally used a cursive hand, while Carla's was a rounded, clear block lettering.

17 Fri   Fog - Rain   Carla went to creek for water, [...] shout & answered about 2½. Sounded like copter near, but saw nothing in air. [...] about 5. Sky this period - ugh. Carla [...] looks like fog will burn off to clear

18 Sat.   Cold [...]

19 Sun.   Awful cold nite. Temp 30 @ 9½. Fog blowing off. Overcast. Carla for water - lost most on way back by slipping. Have heard no planes. Our 9th day. 1 glass jelly left. Getting weak. Fear Al did not make it to help. Making cards out of upholstery to entertain ourselves. Feet need attention, as well as stomachs.

20 Mon   Rain, snow. No change. Feet painful. No trip to creek - collected rain water.

21 Tues   Overcast. Fog. Planes seem to be flying down our valley, tho too high to see us. Have lost all hope. Suggest suicide pills in 1st aid kits. Can't try to walk out - snow too deep for us. Will stay here til gangrene, cold, or malnutrition gets us. Phyllis & Carla struggled to stream - Treacherous trip. Cold & wet on return. Exhausted too weak.

22 Wed   (over)

Foods most wanted:
1 gal [...] or gas
Root Beer
whiskey + 7up
[...]
cola
milk
water to drink
soups
chili - [...] bread + butter
CC dinner
Pot roast
fruit - [...] + apples
Hot chocolate
Hamburgers
french fries
Ice cream - cones + shakes

Baked Pot.
grapes
Dieritos [Fritos] = Popcorn
Pasty
Fish & chips
Oreos
Calif. crunchy cookies
Chinese food
Ham + dill sandwich
Ham
Tacos - Burrito
French dip sandwich
Tuna fish
Rice-a-Roni
Chicken
Spaghetti
Hot Dogs
Macaroni & cheese
hot fudge sundae

pancakes & sausage
Bacon & Eggs
Toast & jam
English Muffins
Peanut Butter & jelly
Sauer Kraut + sausage
Peanut Butter + celery

Notes of despair in the diary, ten days in... By now their feet were failing and the snow was too deep to walk through. Suicide pills were suggested and there were long lists of "dream foods."

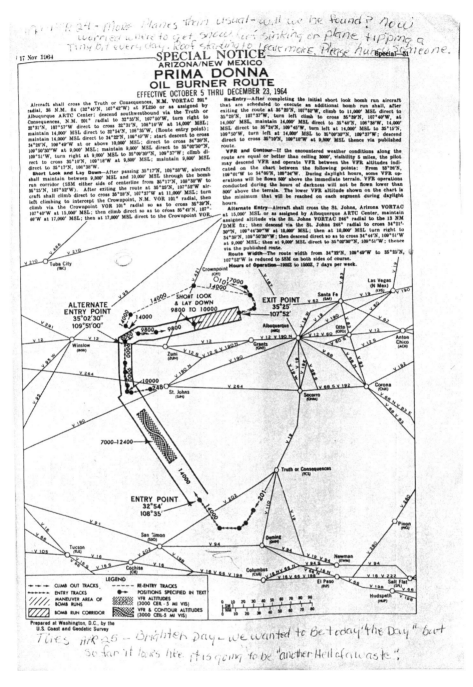

The sad title of the Saturday Evening Post article, "Please Hurry Someone" came from this entry. (The "Oil Burner Routes" were aerial refueling routes for military aircraft.)

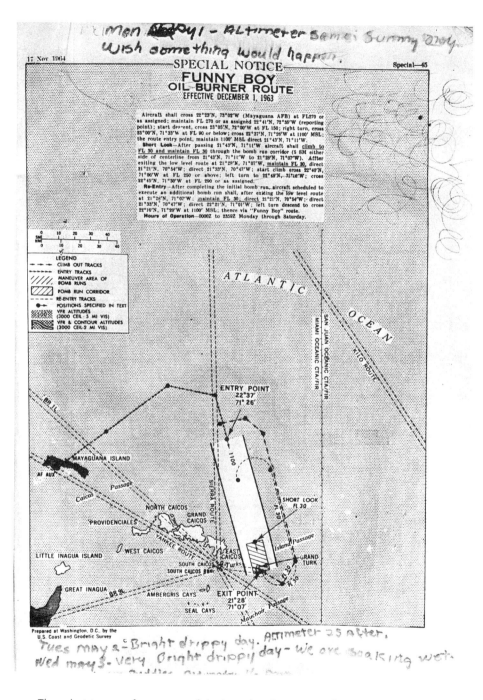

These last two are from some of the later, last few pages of the diary, written around the edges of the charts in text-message-like, brief entries. By the beginning of May, spring came to Shoemaker Bally, but the end was near.

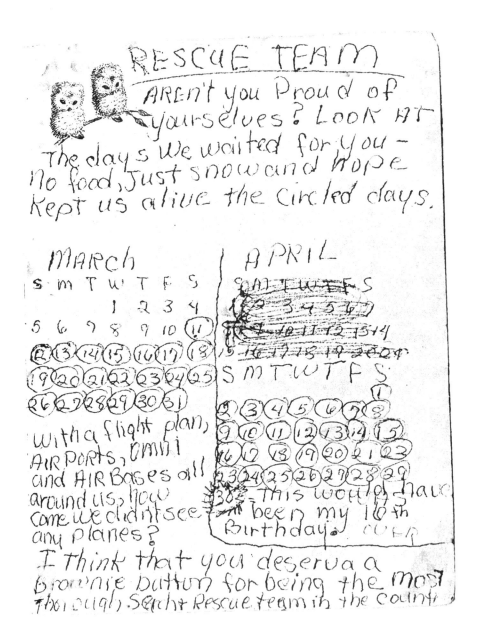

The top sheet found on the diary by Hilda and Floyd Bolling. Though she was long gone, Carla's spirit remained with these haunting words.

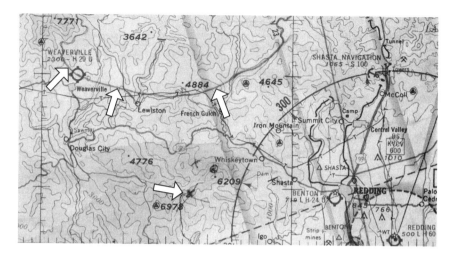

This is a zoom-in of the old chart excerpt shown on Page vi, highlighting where Al, Jr. made a pencil-line indication of the farthest southern extent of his 1967 air search, which shows how close he got to the crash site. He later drew a little airplane symbol showing the exact site of the crash itself…shown about 5 miles farther south from his search-line (see Page 141). The other "X"-mark might be a high-point the brothers used in the ground search, to survey the terrain. The Redding area shows at the bottom right.

The list of Al, Sr.'s personal belongings found on 9/23/69 scattered at his final resting place on the mountain, and turned over to the Redding mortuary who then released them to Al, Jr.

APPENDIX

Senate Record Excerpts

These five excerpts from *The Congressional Record*, four from the Senate and one from the House sessions, make historical account of one Colorado senator's efforts to bring forward legislation regarding what he then referred to as "crash locator beacons" — which are the current emergency locator transmitters (ELTs) aboard every aircraft today.

A pilot himself, Sen. Pete Dominick had been working on several aviation safety issues, and around the time of the Oien's accident he began to focus on mountain flying safety, aviation weather services improvement, and the availability of radiobeacon-type locator devices for airplanes. The record shows him reintroducing the subject in two sessions in August 1967:

> *"This situation can be corrected at relatively low cost. The equipment to help solve it already is available. The legal authority to require its use is already on the statute books. The Federal Aviation Administrator has authority under the Federal Aviation Act to require aircraft to be equipped with such equipment as are shown to be essential to safety. Yet it has not acted to require that general aviation aircraft be equipped with crash locator beacons. Why not?"*

Then when the news broke in October about the Oien's plane with the remains of Carla and Phyllis located, on 10/6/1967 he not only refers to their struggle logged in their diary but requests that it be brought into the *Record* (by reprinting a news article from a Washington, DC paper that carried the story on October 3rd, with entries from the diary itself). Dominick states that

> *"...In most cases of lost aircraft, no record is left of the agony and suffering of the victims, but the picture is made painfully clear by reading of the 7-week ordeal of Mr. and Mrs. Oien and their*

197

daughter. Carla noted in one entry of her diary: 'I hope you are happy, Search and Rescue. You haven't found us yet.' The fault lay not, however, with the searching mission, but quite simply with the fact that they did not have a crash-locator beacon."

Later, in 1969 and 1970, support was building for Sen. Dominick's amendments to a bill being introduced (the "Airport and Airways Development Act of 1969/70"), which would legislate the requirements for ELTs in airplanes. One excerpt for March 1969 shows the support Dominick received in the House as well, via Senator Ottinger of New York (who also brings up the Oien crash). These passages in the *Record* are evident of Dominick's frustration:

"…I pointed out that the FAA had tested five pieces of equipment, that they had approved them, that they had said the equipment was adequate for the purpose, but that they had refused to require the manufacturers to put them in the aircraft. I asked why, and I have received no answer from the FAA."

In the last excerpt from February of 1970, Senators Allott and Goldwater spoke up in enthusiastic support of Dominick's amendment and the record concludes with agreement on the amendment.★ (Thanks are due to Mike McDonald, a long-time CAP member, for kindly obtaining these excerpts for the author to use in this book.)

Later on, Senator Dominick's persistence paid off with a little help from Senator Warren Magnuson of Washington… On December 29, 1970, Public Law 91-596, the law requiring ELT beacons aboard most U.S. aircraft, was signed by President Nixon as a rider to the OSHA Bill. (The ELT law had been inadvertently left unsigned earlier in the year, hence the last-minute attachment to OSHA legislation.)

★The excerpts are taken from PDFs created by and available through HeinOnline.org (online digital content division of William S. Hein & Co., Inc.), are copyright 2013 HeinOnline, and are used with permission. (An example of the original PDF version is shown at right.)

countered in trying to organize its project demonstrate how hard it is for an inexperienced group to get into the housing business.

Some of the problems must still be resolved before the council files its formal application for a mortgage, but Mr. Naclerio expressed confidence that "we've ironed out all the problems." The mortgage would bear only 3 per cent interest.

The council's first idea was to sponsor just housing.

"But every time we went into a building around here to help the tenants we found they had other problems—welfare, education, health," said Pete Velez, supervisor of the council's redevelopment program.

Therefore, when the council drafted plans about two years ago it included a health clinic, a swimming pool, a gymnasium, classrooms and other facilities in several low buildings. The architects, Gerald Silverman and Robert Cika, designed four 25-story apartment houses rising above them.

SITE STILL A SECRET

The council has selected a block in the northern end of East Harlem, which runs from 96th to about 125th Street, and east of Fifth Avenue to the East River. It will not disclose the precise site for fear of pushing up the purchase price.

The most difficult problem was making sure the revenue from the stores and community facilities would contribute enough to the mortgage payments so the council could charge low rents for the apartments. The plan is to rent a two-bedroom apartment for about $93 a month.

The F.H.A. was particularly concerned that the council secure long-term leases from prospective tenants to guarantee a steady income from the non-residential space.

"Everybody cooperated but nobody would go out on a limb," said Robert Meltzer, another council housing aide. "The possible tenants wanted a commitment from F.H.A. before they would sign up, and F.H.A. wanted the leases before they would make a commitment. We ran around in circles for two years."

Mount Sinai Hospital has now agreed to run the health clinic, Mr. Meltzer said, and several universities and city agencies are interested in renting other space.

MAZE OF REGULATIONS

A second major problem was that the council workers had to grope through a maze of regulations that even professional housing experts find baffling.

Such questions as how much tax abatement the city would grant the projects, and whether the area would be designated for urban renewal assistance, have required innumerable meetings, conferences and correspondence.

Council workers and city officials charge each other with dragging their feet, and even at this point neither side knows exactly what has to be done before the project can move ahead.

A third obstacle is the council's lack of money. Architects and accountants must be paid and options must be taken on the land before the F.H.A. mortgage becomes available. Mr. Meltzer estimated the council needed $54,000.

The Legislature passed a bill last year providing $10 million in "seed money" to assist groups like the council. But the state's regulations are so strict that a group must present almost as much evidence to get seed money as to get a mortgage. Housing experts thus consider the fund virtually useless.

Mr. DOMINICK. Mr. President, I suggest the absence of a quorum.

The ACTING PRESIDENT pro tempore. The clerk will call the roll.

The legislative clerk proceeded to call the roll.

Mr. DOMINICK. Mr. President, I ask unanimous consent that the order for the quorum call be rescinded.

The ACTING PRESIDENT pro tempore. Without objection, it is so ordered.

AIR SAFETY

Mr. DOMINICK. Mr. President, recently the news media reported in detail the harrowing episode of the crash between a Piedmont 727 and a private executive airplane, a Cessna 310. Promptly, certain Members of the House raised a cry against general aviation and in favor of banishing them either from the airways entirely or to non-congested fields. This reaction was based on emotion, not on fact. Asheville is so noncongested, that, according to the FAA, it does not rate a radar system. Both planes were on instrument clearances; yet someone cleared the 727 through the very altitude the Cessna was told to maintain. Although both were on instrument clearances, it is not clear whether either was on airways control radar; and if they were not, it is not clear why not.

Having said this much and since I am about to be highly critical of the top management of the FAA and the Weather Bureau, I do want to pay some compliments to FAA flight personnel and to Weather Bureau employees. The FAA personnel at flight service stations, the various flight centers controlling instrument flights, and the tower operators do a truly fabulous job in routing ever-increasing air traffic swiftly and expeditiously and relatively safely—at least in comparison to automobile travel. The Weather Bureau personnel, too, do their best to be helpful, even though their inadequate information creates great scepticism in every weather forecast.

The problems are not with these personnel, but with inferior planning at the top levels. Today, I want to concentrate on general aviation and its problems. Later I will try to outline some ideas for airline aviation. The great Rocky Mountains of the West have contributed their share of accidents, and much of my remarks today will deal with methods to help pilots in this area.

By introduction, I want to say that I have held a flying license for 32 years, have amassed over 5,000 hours of pilot time in single- and multi-engine aircraft, and have flown the Rocky Mountain area for 20 years. Hence, I should have considerable firsthand knowledge of some of the problems involved.

The number of private or general aviation has grown steadily since the end of World War II. There are now over 102,-000 personal and corporate aircraft, compared with the fleet of airline planes numbering only 2,379.

I might say this is a ratio of almost 50 to 1.

During the past 10 years the number of accidents involving general aviation aircraft has likewise grown each year. As the number of accidents has increased, the percentage of fatalities has remained more or less constant at about 10 percent.

For example, in 1957 the 4,200 acci-

dents involving general aviation aircraft produced 438 fatalities. In 1961, there were 4,625 accidents and 426 fatalities. Last year, 5,425 accidents resulted in 538 fatalities. In Colorado alone, from 1964 to 1966, 349 general aviation accidents resulted in 81 fatalities.

But these figures do not tell the complete story. They do not show, for example, that the aircraft accidents which occur more than 5 miles from an airport account for only 20 percent of the total accidents, but result in 53 percent of the fatalities. These are the aircraft accidents which result in tens of thousands of hours being spent by the U.S. Air Force, the Civil Air Patrol, and the Coast Guard searching for the downed aircraft each year.

Since May 29, 1946, through December 31, 1966, FAA records show that the U.S. Air Force flew 1,580,315 hours in search and rescue operations. In fiscal year 1966 alone, the Air Force flew 57,585 hours in search and rescue operations at a total cost to the taxpayers of $59,224,142. The Air Force claims to have saved 525 persons as a result of these operations last year. This amounts to a cost of $112,-808 per person sayed. These figures do not include the costs incurred by the Civil Air Patrol or other Air Force units which often participated in these search and rescue operations.

FAA records show that from February 1, 1961, through December 31, 1966, the Civil Air Patrol flew 104,175 hours on search and rescue missions. In calendar year 1966 members of the Civil Air Patrol flew 21,868 hours on these missions. I do not have any accurate cost figures, as the members of the Civil Air Patrol fly their own private planes and are reimbursed by the Air Force for the fuel and oil used only if they request reimbursement.

There is much that can be done to improve this situation, especially with regard to the 20 percent of the total general aviation accidents which occur more than 5 miles from an airport. Important factors in these accidents have been weather and terrain. When a pilot has inaccurate or incomplete information about either, he is in trouble. If he has poor information about both, he is courting disaster. The lack of such information is not always the pilot's fault. This is especially true regarding weather in certain areas of our Rocky Mountain West.

For years I have been urging the U.S. Weather Bureau to improve its weather reporting service to aviation in the rugged mountainous areas of my own State. Unless you have experienced the weather closing in on you while flying through high mountain passes, you cannot fully appreciate the seriousness of that situation. I have had it happen to me dozens of times and, believe me, it is a frightening and dangerous experience. Most often it happens not because the pilot failed to check the weather reports or disregarded information he was given—it most often happens because the information just is not available on a current enough basis, or in many cases is not available at all.

I might interpolate here to say that in

Congressional Record—Senate, August 7, 1967; pages 21633 through 21635

AIR SAFETY

Mr. DOMINICK. Mr. President, recently the news media reported in detail the harrowing episode of the crash between a Piedmont 727 and a private executive airplane, a Cessna 310. Promptly, certain Members of the House raised a cry against general aviation and in favor of banishing them either from the airways entirely or to noncongested fields. This reaction was based on emotion, not on fact. Asheville is so noncongested, that, according to the FAA, it does not rate a radar system. Both planes were on instrument clearances; yet someone cleared the 727 through the very altitude the Cessna was told to maintain. Although both were on instrument clearances, it is not clear whether either was on airways control radar; and if they were not, it is not clear why not.

Having said this much and since I am about to be highly critical of the top management of the FAA and the Weather Bureau, I do want to pay some compliments to FAA flight personnel and to Weather Bureau employees. The FAA personnel at flight service stations, the various flight centers controlling instrument flights, and the tower operators do a truly fabulous job in routing ever-increasing air traffic swiftly and expeditiously and relatively safely—at least in comparison to automobile travel. The Weather Bureau personnel, too, do their best to be helpful, even though their inadequate information creates great scepticism in every weather forecast.

The problems are not with these personnel, but with inferior planning at the top levels. Today, I want to concentrate on general aviation and its problems. Later I will try to outline some ideas for airline aviation. The great Rocky Mountains of the West have contributed their share of accidents, and much of my remarks today will deal with methods to help pilots in this area.

By introduction, I want to say that I have held a flying license for 32 years, have amassed over 5,000 hours of pilot time in single- and multi-engine aircraft, and have flown the Rocky Mountain area for 20 years. Hence, I should have considerable firsthand knowledge of some of the problems involved.

The number of private or general aviation has grown steadily since the end of World War II. There are now over 102,-000 personal and corporate aircraft, compared with the fleet of airline planes numbering only 2,379.

I might say this is a ratio of almost 50 to 1.

During the past 10 years the number of accidents involving general aviation aircraft has likewise grown each year. As the number of accidents has increased, the percentage of fatalities has remained more or less constant at about 10 percent.

For example, in 1957 the 4,200 accidents involving general aviation aircraft produced 438 fatalities. In 1961, there were 4,625 accidents and 426 fatalities. Last year, 5,425 accidents resulted in 538 fatalities. In Colorado alone, from 1964 to 1966, 349 general aviation accidents resulted in 81 fatalities.

But these figures do not tell the complete story. They do not show, for example, that the aircraft accidents which occur more than 5 miles from an airport account for only 20 percent of the total accidents, but result in 53 percent of the fatalities. These are the aircraft accidents which result in tens of thousands of hours being spent by the U.S. Air Force, the Civil Air Patrol, and the Coast Guard searching for the downed aircraft each year.

Since May 29, 1946, through December 31, 1966, FAA records show that the U.S. Air Force flew 1,580,315 hours in search and rescue operations. In fiscal year 1966 alone, the Air Force flew 57,585 hours in search and rescue operations at a total cost to the taxpayers of $59,224,142. The Air Force claims to have saved 525 persons as a result of these operations last year. This amounts to a cost of $112,-808 per person saved. These figures do not include the costs incurred by the Civil Air Patrol or other Air Force units which often participated in these search and rescue operations.

FAA records show that from February 1, 1961, through December 31, 1966, the Civil Air Patrol flew 104,175 hours on search and rescue missions. In calendar year 1966 members of the Civil Air Patrol flew 21,868 hours on these missions. I do not have any accurate cost figures, as the members of the Civil Air Patrol fly their own private planes and are reimbursed by the Air Force for the fuel and oil used only if they request reimbursement.

There is much that can be done to improve this situation, especially with regard to the 20 percent of the total general aviation accidents which occur more than 5 miles from an airport. Important factors in these accidents have been weather and terrain. When a pilot has inaccurate or incomplete information about either, he is in trouble. If he has poor information about both, he is courting disaster. The lack of such information is not always the pilot's fault. This is especially true regarding weather in certain areas of our Rocky Mountain West.

For years I have been urging the U.S. Weather Bureau to improve its weather reporting service to aviation in the rugged mountainous areas of my own State. Unless you have experienced the weather closing in on you while flying through high mountain passes, you cannot fully appreciate the seriousness of that situation. I have had it happen to me dozens of times and, believe me, it is a frightening and dangerous experience. Most often it happens not because the pilot failed to check the weather reports or disregarded information he was given—it most often happens because the information just is not available on a current enough basis, or in many cases is not available at all.

I might interpolate here to say that in our particular State we have over 50 mountains which exceed 14,000 feet in height. When one is flying a small airplane, the degree of wind turbulence in those mountains can cause the situation to become extremely perilous. If the pilot were flying in a flat area he could readily see what the weather turbulence might be, simply by observation.

Last year the Weather Bureau assured me that decided improvement in weather reporting could be brought about in Colorado for the relatively low cost of $43,855—absolutely a peanut within the budget before us this year and last year. They outlined proposed changes as follows:

First. Five of the weather reporting stations would be expanded to provide 3-hourly observations for 24 hours each day by contract observers, and their hours for reporting would be increased as follows:

Montrose and Gunnison, increase from 14 to 18 hours.

Salida, increase from 6 to 11 hours.

Aspen, increase from 10 to 18 hours.

Durango, increase from 19 to 24 hours.

Second. Two new weather reporting stations would be added at Nucla and Walden with 3-hourly observations for 24 hours each day, and reporting 8 hours a day.

Third. The Weather Bureau station at Alamosa would be expanded to 3-hourly observations for a full 24 hours each day, and reporting services would be increased from 17 to 19 hours a day.

Annual cost of increased observations, $32,855.

Annual cost of communications, $11,000.

Total annual cost of revised observing program for Colorado, $43,855.

I was assured this item could be included in the Weather Bureau's budget for fiscal year 1968. I urged unsuccessfully last year that in view of the potential benefits compared to the small cost, it be done sooner. The improvements have not been accomplished, and we now have received the 1968 budget. This item is not included. It is only fair to ask, Why not?

When you look at the cost incurred by the Air Force alone for each person it saves through search and rescue operations, the cost appears very small for improved weather reporting to help prevent aircraft accidents—as well as providing better background information for more accurate weather forecasting for all people. Remember that the Air Force alone reported that it spent $112,808 during 1966 for each person it saved. The Weather Bureau appears unwilling to spend a third of that amount to help prevent these accidents.

There are other improvements which could be made which would result in saving more lives and reducing costs in search and rescue operations following aircraft accidents. In a substantial number of cases, the searching aircraft have only a vague idea of where they should begin to search for the missing plane. Weather and terrain again are important factors which hamper search and rescue operations.

National figures for 1965 show 22 aircraft with 37 persons aboard lost in the most literal sense of the word. Although diligent searches were made, not a trace has yet been found of these people or the missing planes. No one knows how many of them may have been overflown many times and not seen from the air. I know this has happened in dozens of cases which have occurred in the Rocky Mountain area. The loss of Dr. Lovelace and his family last year is a recent example of that situation. In that case, I talked to many people who were in the search and rescue operation, they told me they went over the spot where the airplane went down seven times before they happened to find it. The tragedy of the story is that Dr. Lovelace and the passengers in that airplane were alive when they went down on the snow- and tree-covered slope outside of Aspen. They died because we could not find them in time.

This situation can be corrected at relatively low cost. The equipment to help solve it already is available. The legal authority to require its use is already on the statute books. The Federal Aviation Administrator has authority under the Federal Aviation Act to require aircraft to be equipped with such equipment as are shown to be essential to safety. Yet it has not acted to require that general aviation aircraft be equipped with crash locator beacons. Why not?

As long ago as January 9, 1964, FAA issued Advisory Circular No. AC 170-4 which stated as follows:

1. *Purpose.* This circular informs the aviation community and industry of recent FAA activities concerning the use of crash locator beacon systems and their respective potential application as an emergency signaling device.

2. *Background.*

(a) An evaluation was recently conducted by the FAA in the Los Angeles and Salt Lake City areas to determine the effectiveness of such a system by using three individual and commercially developed equipments. Specific equipments used during the test program do no represent all available types. * * * Simulated test conditions were exercised in geographical areas including mountains, flat, and congested (radio signal-wise) areas.

(b) *Results of the test program clearly indicated that such a beacon can successfully radiate energy to permit a suitably equipped search aircraft to identify and "home-in" on the transmitted signal.*

A chart included on page 2 of the circular shows that the signal from one of these crash locator beacons can be received by a plane flying at 10,000 feet within a range of 70 nautical miles. The FAA, in a letter to me dated May 12, 1967, in response to my inquiry, stated in part as follows:

Crash locator beacons have been perfected, selling at prices exceeding $200. The question is: Should FAA require their use by regulation? We do not recommend such regulatory action.

Air Force, Navy and Coast Guard planes all have crash-locator beacons installed as standard equipment, as do commercial aircraft engaged in certain types of air carrier operations. Why not require their use on general aviation aircraft? The savings in lives of searchers as well as victims of air crashes would be significant.

I repeat, the FAA says that they have been perfected, and the question was, Should their use be required by regulation? The FAA, without any explanation of any kind, simply said:

We do not recommend such regulatory action.

Mr. President, in an audit report from the General Accounting Office, sent to Congress May 26, 1967, the Comptroller General concluded that—

The Federal Aviation Administration in the Pacific Region had procured goods and services costing $267,000 which could not be justified or could be justified only partially by the conditions existing at the time of the procurements or by the benefits received by the Government.

Among the findings in the report:

The Pacific region in 1964 acquired an alarm/sound system for its new headquarters building, leased for a 10-year period at $10,600 per year. The report states:

Because alarm systems that might be needed had already been provided by the lessor of the building or by the State, we believe that the principal purpose being served by the FAA alarm/sound system was to provide the less essential features of music and paging.

The lease, incidentally, has been terminated since the report was issued, but the system had already cost the Government about $36,000.

The report continues:

Also in 1964, the Pacific Region purchased 148 clothes dryers at a cost of about $12,500 for installation in employee housing on Wake Island. Additional costs of about $4,700 were incurred for shipping charges. About one year after the purchase of the dryers, the Pacific Region issued a contract to provide wiring, plumbing, and installation and modification work to make the dryers operational. The cost of this work totaled about $25,000. In our opinion, the purchase of dryers under these circumstances was premature and evidenced a lack of prudent management of Government resources.

Several other expenditures are detailed, such as procurements totaling about $27,000 for a facility which was about to be phased out of service; costs of over $30,000 for dedication ceremonies for three FAA facilities; $5,300 for audiometer booths which were subsequently placed in storage for 2 years for lack of space, then later declared surplus when the island was decommissioned; purchase of mobile radio equipment which was never installed because the cars were already equipped with mobile telephones.

It seems unfortunate, indeed, that such unjustified expenditures are made by a Government agency, in view of the present rate of deficit spending by the administration; but it is even more unfortunate in view of the critical need for research and development in the field of air safety. The Members of Congress are told time and again that for improved air safety equipment, more appropriations are needed. Yet, money is wasted on items such as those listed above while the FAA flatly rejects air safety devices which cost the Government nothing and which would save the lives of many people and save the taxpayers enormous amounts in air rescue operations. With more economical procurement policies than the ones adopted in the Pacific region, the FAA could have applied at least $267,000 toward further refinement of the many air safety devices and procedures currently under study.

A letter to the editor of Aero West magazine, May 1967, issue, eloquently states the case for requiring crash locator beacons on all aircraft. I read the letter in full because it is as dramatic an example as can be stated. It reads as follows:

You can remove my husband's name from your sample copy mailing list. He was killed in 1965.

I note in your editorial on page 6, and in the article on Senator Dominick's talk, that it would be possible to install an electronic marker beacon in the tail of every plane. If such equipment is available and has proved its effectiveness, it is unthinkable that it should not be required. Such equipment could have saved a week of fruitless searching for the plane in which my husband crashed with three passengers.

The crash occurred only about 33 miles from our home airport, on a heavily timbered mesa. Searchers estimated they flew over the spot at least 20 times but could not see a sign of the plane because of the way it had dived into the timber.

Even when ground vehicles pointed out the crash after it had been discovered, air searchers could not see it.

The tail was relatively undamaged, and though it could not have saved these men, who were all killed on impact, it could have shortened the hazardous, expensive search conducted by the Civil Air Patrol and the local pilots.

The idea that such a safety device would make people nervous about flying, or scare off buyers, is too absurd. If the device is very costly and cumbersome, it should be improved, not ignored. But it should most certainly be standard equipment.

That letter was from one of my constituents in Gunnison, Colo.

Mr. President, I state unequivocally that the failure of the Weather Bureau to install proper weather monitoring stations in the mountains of Colorado has meant the death of many fliers. I state unequivocally that the continued failure to install these stations and to man them with competent observers and reporters will mean more deaths. I state that the failure of the FAA to require the installation of locator beacons resulted in the death of Dr. Lovelace and his family, the loss of many other fine citizens, and has literally cost the taxpayers millions of dollars.

Mr. President, we need action now, not studies, not excuses, not delays, but ac-

tion, if we are to save human lives.

The ACTING PRESIDENT pro tempore. Under the previous unanimous consent agreement, the Senator from Ohio is recognized.

Congressional Record—Senate, August 18, 1967; pages 23272 through 23275

GENERAL AVIATION SAFETY

Mr. DOMINICK. Mr. President, recently I delivered a speech on general aviation safety and steps which I thought should be taken to promote this safety, particularly in our own Rocky Mountain area.

One such proposal was to implement a program already developed by the Weather Bureau, but which had not been funded, despite their recommendation that it go into effect. As a matter of fact, they apparently had not pushed for it in the budget; it was not included in the budget.

However, I had an opportunity to appear before the very distinguished Appropriations Subcommittee on the Departments of State, Commerce, Judiciary, and related agencies, to push this program, at a minimum cost of $43,855. I received a very good reception before the subcommittee, and both the Denver Post and the Rocky Mountain News have given good coverage and support to my effort.

At this point I ask unanimous consent to have printed in the RECORD a detailed article that appeared in the Denver Post of August 13, written by Dan Partner, a very fine reporter, and two fine articles dated August 19, one by Jim Foster of the Rocky Mountain News and one by Barnet Nover of the Denver Post.

There being no objection, the articles were ordered to be printed in the RECORD, as follows:

SENATOR DOMINICK: WEATHER PLAN FUNDS URGED

(By Barnet Nover)

WASHINGTON.—Senator Peter Dominick, R-Colo., urged the Senate Appropriations Committee Tuesday to include in its appropriations for the Department of Commerce an item of $43,855 for improved weather observation in Colorado.

"From 1964 to 1966 in Colorado alone," he said, "349 general aviation accidents resulted in 81 fatalities. Many of these accidents were caused by weather closing in on the pilot. Most often this happens because complete weather information is not available on a current enough basis, or is not available at all."

Dominick told the committee that his proposals for greater air safety had aroused enthusiastic support.

"Anyone who has flown in the mountains knows how critical weather information can be."

Dominick also stressed that the amount asked for improved weather information is less than the average amount that was spent by the Air Force—$112,808—for each person saved through its search and rescue operations in fiscal 1966.

The money would provide for expanded weather observations at five existing stations and provide for two new reporting stations at Nucla and Walden.

DOMINICK'S AIR SAFETY PLAN WELL RECEIVED

(By James Forter)

WASHINGTON, Aug. 15.—Recommendations by Sen. Peter H. Dominick to improve air safety may become a guide for nationwide upgrading of weather observation, The Rocky Mountain News learned Tuesday.

Immediately after hearing testimony by the Colorado Republican—an active pilot with more than 5000 hours of flying time—a Senate appropriations subcommittee called for the Environmental Sciences Services Administration (ESSA) to respond with comments and specific recommendations.

Subcommittee Chairman John McClellan (D-Ark.), obviously impressed with Dominick's experience in flying over Colorado's mountains, directed that the senator's full testimony be sent to ESSA Administrator Robert W. White.

McClellan suggested that Dr. White might want to testify before the subcommittee.

McClellan said he thought Sen. Dominick's ideas were feasible and that perhaps they could be used as a guide for establishing better weather service throughout the country, particularly in mountainous areas.

Dominick noted that his Aug. 7 Senate speech on air safety drew favorable reaction from pilots and flying enthusiasts across the country.

In a letter to the 14 members of the state, commerce and judiciary subcommittee, Dominick pointed out that despite assurances from the Weather Bureau, provisions for improved weather observation were left out of the 1968 budget.

Dominick's 3-point plan would include:

Five weather stations would be expanded to provide 3-hourly observations and longer hours of reporting—Montrose and Gunnison increased from 14 to 18 hours, Salida increased from six to 11 hours, Aspen increased from 10 to 18 hours and Durango increased from 19 to 24 hours.

Two new weather stations would be added at Nucla and Walden with 3-hour observations 24 hours a day and reporting eight hours a day.

Alamosa services would be expanded to a full 24-hour observation day with reporting increased from 17 to 19 hours a day.

Dominick testified that the annual cost of expanded service would be only $43,855. He pointed out that the Air Force spent an estimated $112,808 for each person saved through search and rescue operations in fiscal year 1966.

"The sum of $43,855 seems very small indeed for improved weather reporting to help prevent aircraft accidents as well as to provide better background information for more weather forecasting for all people," Dominick declared.

[From the Denver Post, Aug. 13, 1967]

OFFICIALS AGREE WITH DOMINICK'S AIR SAFETY STAND: LACK OF FUNDS—NOT DESIRE—SLOWS ACTION

(By Dan Partner)

Officials of two U.S. agencies attacked by Sen. Peter Dominick on the Senate floor last week have no quarrel with the Colorado Republican. But they point out that the lack of dollars—not desire—prohibits the action he demands.

Dominick was unusually bitter in his criticism of the U.S. Weather Bureau and the Federal Aviation Administration (FAA) in regard to flying safety. He said the failure of the Weather Bureau to install proper weather monitoring stations in the Colorado mountains has "meant the death of many flyers." The failure of the FAA to require the installation of crash locator beacons has resulted in loss of lives, hazardous searching operations, and has cost taxpayers millions of dollars, he said.

Dominick quoted a figure of $43,855 as the Weather Bureau's estimate last year for an improved weather observing program for Colorado. The proposed changes included new weather reporting stations at Nucla and Walden and expansion of operations at Montrose, Gunnison, Salida, Aspen, Durango and Alamosa.

BUREAU FAILURE

The failure of the bureau to include the project in its 1968 fiscal year budget ignited the blast from Dominick in his Senate speech.

"The Air Force alone has incurred a cost of $112,808 for each person saved through search and rescue operations, but the Weather Bureau appears unwilling to spend a third of that amount to help prevent these accidents," Dominick said.

A spokesman for the Environmental Science Services Administration (ESSA), which includes the Weather Bureau, said Saturday the Colorado requirements were "valid" but that no funds were available "at present." He said the project definitely would be budgeted in fiscal 1969.

"We recognize the requirements in Colorado and there are similar requirements in other Western states and in other parts of the country," said John Eberly, ESSA executive officer. "We're doing everything we can with the resources available."

Dominick told The Denver Post Saturday he was pushing for the $43,855 in the 1968 budget because Colorado "has the worst weather reporting system in the Rocky Mountain area." He said he was preparing a letter to the subcommittee on appropriations urging addition of the money. Failing this, Dominick will add the amount when the appropriations bill is introduced in the Senate. Meanwhile, he's attempting to get the job done with existing funds.

EXPERIENCED PILOT

Dominick has logged more than 5,000 hours as a pilot and has flown in the Rocky Mountain area for 20 years. He's had several frightening experiences due to bad weather forming unexpectedly and he is convinced improved weather reporting can prevent such situations.

Regarding the failure of the FAA to require general aviation aircraft to be equipped with crash locator beacons, Dominick said: "National figures for 1965 show 22 aircraft with 37 persons aboard lost in the most literal sense of the word. Although diligent searches were made, not a trace has yet been found of these people or the missing planes.

"No one knows how many of them may have been overflown many times and not seen from the air. I know this has happened in dozens of cases which have occurred in the Rocky Mountain area. The loss of Dr. Lovelace and his family is a fairly recent example of just that situation happening."

(Dr. W. Randolph Lovelace, his wife and the pilot of their rented plane died following a crash near Aspen, Colo., in December 1965.)

Dominick contends crashed planes can be located at "relatively" low cost.

"The equipment to help solve it already is available," he said. "The legal authority to require its use is already on the statute books. The FAA has authority to require aircraft to be equipped with such equipment as are shown to be essential to safety. Yet it has not acted to require that general aviation aircraft be equipped with crash locator beacons. Why?"

A spokesman for the FAA in Washington told The Post that, after the Lovelace accident, the agency published its intent to require all air taxi operators to equip their planes with crash locator beacons. The beacons are radio transmitters that emit a signal to be picked up by search aircraft.

"We were deluged by objections to the plan," the spokesman said. "The plan was suspended."

The FAA favors equipping planes with beacons but has received "luke warm" response from private pilots and flying organizations. It was noted that the beacons cost from $200 to $500 each and that only a small percentage of the nation's private pilots fly over terrain that would require the equipment in locating a crash.

"Most pilots don't want to spend that kind of money on equipment they don't think they'll ever need," the spokesman said. "Most of them don't even spend money for a first aid kit."

Dominick, however, believes pilots should be required to install the equipment to save taxpayers' money used in search and rescue operations. He said the Air Force flew 1,580,316 hours in search and rescue missions between May 29, 1946, and Dec. 31, 1966. During the 1966 fiscal year, such missions required 57,585 hours and cost $9,224,142, he said.

"The Air Force claims to have saved 525 persons as a result of these operations last year," Dominick said. "This amounts to a cost of $112,808 per person saved. These figures do not include the costs incurred by the Civil Air Patrol or other Air Force units which often participated in these operations."

. . .

Mr. DOMINICK. It is my hope, Mr. President, that continued interest in this problem will bring more adequate weather observations, more adequate weather communications, and more accurate weather forecasting, thus resulting in general aviation safety and benefit to the people as a whole.

Ironically, at the very time I was advocating better weather reporting and crash locater beacons, two general aviation accidents occurred in our mountains in Colorado. One of these crashes involved a small plane apparently caught in the very bad downdrafts that happen from time to time from the wind pouring over the Continental Divide. This accident probably could have been avoided, and it is probable that more adequate weather reporting could have persuaded the pilot to have taken another route where he could have avoided this turbulence.

I ask unanimous consent that an editorial with respect to this occurrence, which appeared in the Denver Post of August 15, be printed at this point in the RECORD.

There being no objection, the editorial was ordered to be printed in the RECORD, as follows:

A PLEA FOR MOUNTAIN AIR SAFETY

We can't tell precisely how the four people from Missouri met misfortune last week in an airplane crash near Boulder but the incident does prompt us to underscore once again the danger of Colorado flying if you're inexperienced in mountain flying.

The Missouri pilot, who may suffer crippling injuries, left the Longmont Municipal Airport at 10:35 a.m. Thursday en route to Las Vegas, Nev., with his wife and another couple aboard. They had been given three alternate routes by Longmont airport personnel: by way of Laramie, Wyo., Albuquerque, N.M., or over Corona Pass.

Circumstances of their crash near Tolland are not fully known, but professional pilots who know the specifications of the light plane they flew are apt to say: "It's lucky they got as far as they did." The Missourians appear to have been trying to negotiate Corona Pass or one of the other passes in the range west of Boulder.

The point is that the odds were stacked against them. The plane they were flying is a perfectly good plane for the purposes it was intended. But it doesn't have the safety margin at Front Range altitudes with the loading that existed Thursday to cope with the downdrafts frequently found at midday on the east slope of the mountains.

Evidence at the Longmont airport suggests that somebody told the Missouri party—in general conversation—about the dangers of the Front Range. At the same time, we hope whoever gave them the alternative routing via the mountains accompanied that information with the strongest possible warning against such a flight plan.

Largely, the problem remains one of education. Mountain flying in Colorado and other Western states presents hazards which pilots unfamiliar with mountain conditions cannot be expected to handle confidently.

In this regard it is noteworthy that the Federal Aviation Administration is sending out letters to pilots in Colorado, Wyoming and adjoining states urging a "safe holiday weekend" during the forthcoming Labor Day observance.

Safety tips will be given in a small handbill accompanying the letter. Additionally, FAA personnel will donate their own time during the holiday period to helping pilots have safe trips. And the FAA is telling its weather operators to close their telephone reports to pilots with a message about safety.

We're glad to see FAA taking a stronger hand. As more people turn to civilian flying the need for improved flight education grows apace.

Mr. DOMINICK. Mr. President, it will be recalled that I spoke of the need for crash locater beacons in general aviation aircraft. I pointed out that the FAA had tested five pieces of equipment, that they had approved them, that they had said the equipment was adequate for the purpose, but that they had refused to require the manufacturers to put them in the aircraft. I asked why, and I have received no answer from the FAA.

Just to point up this problem, recently an Aero Commander, a very fine airplane, went down in our mountains in Colorado. It was flown by an experienced mountain pilot—at night—and when the crash was finally located, it happened to be very near a place at which I used to land my own airplane in order to go fishing. All four men in the airplane were dead. The bodies of three of them were still in the airplane, and the body of the fourth had been thrown out. All four were fine people from my home county in Colorado.

Before this airplane was reported missing the pilot had filed a flight plan indicating where he was going. Since the pilot later failed to report in, a search pattern was begun. When the plane was found, it was on course, or relatively near course. I want to emphasize that it took an entire week—an entire week of rescue efforts on the ground and in the air, at enormous expense to everybody and with tremendous turmoil to the family of these people—to find the airplane.

I have in my hand an article that appeared in the Rocky Mountain News of August 16, entitled "Relatives Offer Reward in Plane Search."

Frankly, we did not need a reward. People were searching as hard as they could for this airplane, but the search continued for an entire week. If there had been a crash locator beacon in that plane, it could have been located in a

matter of hours; and the expense of the search would be minimized. If any of the men survived the crash, they might have been rescued and their lives saved.

. . .

Mr. DOMINICK. Mr. President, I do not know what it takes to make the Federal Aviation Administration wake up. I do not know whether we have to say to them, "You are not going to get any appropriations unless you do something of this kind." I do not know what it takes to make the policy leaders down there require a simple thing which does not cost the Government any money and which could save millions of dollars in rescue and search efforts and save many lives in the process.

Mr. President, I intend to make a vendetta out of this matter, and I shall keep on and on in connection with every accident I can find until the Federal Aviation Administration decides to do something that should have been done years ago. This step can be taken at no expense to the taxpayer; it can be done with a simple FAA order; it can be done with no real objection from general aviation; and it can be done with the consent of the flying clubs throughout the Nation. However, for reasons I hesitate to mention, apparently the FAA is unwilling to put in the simple requirement for a piece of equipment which will send out a signal so that any plane looking for a downed aircraft can tune in on that beacon, and home in on it. I will point out that the signal from this beacon can be received by another aircraft flying at 10,000 feet within a radius of 70 miles.

Mr. President, it seems strange that we cannot get more than complete silence from the Federal Aviation Administration. Many of the regional men, many of the traffic control people, many of the aviation safety people, and many people who are employees of the Federal Aviation Administration have been writing to me and saying that this is a tremendous proposal and they hope it will be pushed with those who control policy. Still the FAA says "No."

Mr. President, if the only way I can get any reaction from the Federal Aviation Administration in an affirmative manner is to continue attacking them—when I think they have done a very fine job in many other areas—I shall be forced to do that. If the only way I can get them to listen is to try to get cuts in their appropriations so that they will not be spending funds on unnecessary items and will be spending funds on the safety items, I shall take that route. I shall do whatever is needed. I fully intend to continue to try to do something for the people who are running into these problems day in and day out throughout our country.

Mr. President, I ask unanimous consent to have printed in the RECORD a telegram addressed to me from Larry A. Ulrich, Combs Aircraft, Inc., Stapleton Airfield, Denver, Colo.

There being no objection, the telegram was ordered to be printed in the RECORD, as follows:

> DENVER, COLO.,
> *August 15, 1967.*
> Senator PETER H. DOMINICK,
> *Washington, D.C.:*
> Thanks from all of us who fly for your brilliant, forceful and informative Senate speech on weather bureau services and locator beacons. Based on phone calls I've received from FAA and newspapers you may have started things moving. Coincidentally an Aero Commander lost last Wednesday somewhere between Grand Junction and Denver with 4 people aboard; search futile so far, details follow.
> COMBS AIRCRAFT, INC.,
> LARRY A. ULRICH.

Mr. BYRD of West Virginia. Mr. President, I suggest the absence of a quorum.

The PRESIDING OFFICER. The clerk will call the roll.

The legislative clerk proceeded to call the roll.

Mr. STENNIS. Mr. President, I ask unanimous consent that the order for the quorum call be rescinded.

The PRESIDING OFFICER. Without objection, it is so ordered.

Congressional Record—Senate,
October 6, 1967; pages 28107
through 28110

CRASH-LOCATOR BEACONS

Mr. DOMINICK. Mr. President, an article in the Washington Post of 3 days ago was undoubtedly noted by many of my distinguished colleagues with a great deal of remorse. I refer to the report of the diary written by 16-year-old Carla Corbus and her mother, whose remains were found almost 6 months after their plane went down in the mountains of California. I only hope the top echelon of the Federal Aviation Administration, who have consistently refused to require installation of crash-locator beacons in private aircraft, will take note of this tragic incident.

At this point, Mr. President, I would like to recapitulate on some of the points I have brought out in previous statements urging requirement of crash-loca-

tor beacons. On August 7, in this Chamber, I pointed out the number of fatalities resulting from general aviation accidents, noting that 53 percent of these fatalities have resulted from crashes occurring more than 5 miles from an airport—only 20 percent of the total general aviation accidents.

The costs incurred in searching for lost aircraft have been tremendous—$59,224,142 for search and rescue missions flown by the Air Force alone in fiscal 1966. I am sure no one will argue the justification of using taxpayers' funds for these missions, but I do question the refusal of the FAA to take such an urgently needed step to reduce the hazards and the expense of conducting searches for missing aircraft. And, as so tragically illustrated by the report of 3 days ago, crash-locator beacons would most assuredly save lives.

Just 2 months ago, a plane crash in the Colorado mountains was the object of a weeklong search by the Civil Air Patrol and private citizens. I reported in detail to the Senate on this incident in a statement August 18. In this case, there may not have been survivors of the crash—a question which will never be answered. Nevertheless, a full week was spent by ground and air units before the wreckage was finally found, and with no small hazard to pilots trying to fly low enough in mountainous terrain to spot the plane.

In the letters I have received subsequent to my statements calling for requirement of crash-locator beacons, wholehearted support for the proposal has been expressed by pilots in many areas of the country. The few objections raised have been on the grounds of undue expense to owners of private aircraft and the thought that the device would be helpful only for mountain flying.

On the first point, I sympathize with the reluctance of pilots to stand the $200 expense; but for a device that could very well save their own lives, I certainly do not think it would be asking too much. Since it is for their own protection, it seems just a little unreasonable to leave the entire burden to the taxpayers and the Civil Air Patrol when an alternative is so readily available. One FAA representative commented not long ago that there is no law requiring the Air Force, the Civil Air Patrol, or private pilots to undertake searching missions for downed aircraft. No law, perhaps, other than a moral obligation to save human life, regardless of expense, if there is a remote chance of success. Just what does it take to convince the FAA that the possibilities of success could be greatly increased by this one regulation? Should not the burden of the expense rest with those whose lives might be saved through precautionary measures, rather than in expensive, hazardous assessments on those with the moral responsibility to find them, even if it may be too late?

The second objection—that crash-locator beacons would be necessary only in a few areas of the country—I think is a highly fallacious assumption. I doubt that there are many pilots in this country who have not at one time or another found themselves flying over a relatively unpopulated area, or over rough, wooded terrain, or even over a mountainous region. I would stress again the letter I quoted in my statement before the Senate on August 7, from a Colorado woman whose husband was killed in a plane crash near Gunnison, Colo. After a week of searching, the plane was finally located only 33 miles from the airport on a heavily timbered mesa. The plane was not high on a mountain peak, inaccessible to ground searching crews. It was, in fact, finally located by ground crews, and even then could not be spotted from the air because of the way it had dived into the timber. There are many other regions in the United States where a plane could go down and not be spotted from the air for just such a reason. I received a very simple response to my proposal for crash-locator beacons recently, which certainly attests to this fact. It reads:

A few years ago, I crashed on take-off in the mountains of Maine. I wasn't found for 24 hours. With a locator beacon I believe I would have been found very quickly. I favor this recommendation.

Another letter from Ogdensburg, N.Y., reads as follows:

I have read the copy of your recent aviation speech. In my short aviation life (200 hours) I have already learned that the weather station forecasts must be confirmed by personal observation. On one occasion I found the weather report to be quite erroneous only 30 miles from the station. The men working the stations have always spared no effort to assist me, you understand, but I would agree with your thinking on this matter.

I too have spent long hours on a Search Mission in the Civil Air Patrol. I shouldn't have been there, really, as the weather was marginal and I am strictly VFR. However, the thought that people might be alive and lying out in the snow with broken legs kept me going. We never found the wreck. A farmer on a tractor did after we had searched for almost three days. A crash locator beacon would have shown us the way in 30 minutes. Such beacons would be a boon.

This letter is from a doctor in Ogdensburg, N.Y. His experience clearly indicates that there are circumstances other than those to be encountered in our Colorado mountains that make search-and-rescue missions both hazardous and expensive. Ogdensburg, to my knowledge, is not in a mountainous region. It is on the St. Lawrence River, and, I believe, in a relatively flat area; but in this instance the weather conditions hampered search operations, not the terrain.

In other correspondence from the State of New York, I find that there are problems at least comparable to our own Rocky Mountains in the West. In a letter from Col. John C. Campbell, Jr., commander of the New York Air National Guard at Schenectady County Airport, I received the following report as evidence of the need for crash-locator beacons:

I have read your speech to the United States Senate with regard to aircraft crashes. I am in complete agreement with your views. I, too, am strongly in favor of crash locator beacons.

This year at least three private aircraft crashed and were missing in our area. These aircraft crashed in mountainous, wooded terrain and could not be spotted from the air, since the woods had complete leaf cover. All were reported missing for many, many days. Help could not reach them in time, even though the crashes occurred relatively close to high population centers with all the elaborate paraphernalia of our advanced society.

In each case, crash locator beacons would have summoned aid days earlier and would have saved thousands of wasted manhours in futile search.

Over the years, I can recall many missing aircraft crashes in the Catskill, Adirondack, and Green Mountain areas which surround our Capitol District. When a small aircraft crashes in late spring or summertime, the common saying here is to wait for some deer hunter to find the wreck in the fall of the year. A sad commentary, but so very true in too many cases.

In subsequent correspondence with Colonel Campbell, I received further details on five crashes which occurred this year in that region. I ask unanimous consent that Colonel Campbell's letter of September 15 and the attached report from Mr. Francis Mosher, Jr., air safety investigator for the New York office of the Department of Transportation, be printed in the RECORD at this point.

There being no objection, the letter and report were ordered to be printed in the RECORD, as follows:

NEW YORK AIR NATIONAL GUARD,
Schenectady County Airport, N.Y.,
September 15, 1967.
Senator PETER H. DOMINICK,
U.S. Senate,
Washington, D.C.
DEAR SENATOR DOMINICK: I refer to your letter of September 6th and my answering letter of September 15th. The inclosed De-

partment of Transportation letter has the details of the aircraft crashes which have occurred this year in the Albany, New York area.

You will notice that all aircraft were lost for several days, at least. One has not been found yet and one was missing for almost two years. It seems obvious that had crash locator beacons been installed they would have, first, pinpointed these crashes and thus directed rescue efforts to the scene with some dispatch and, secondly, saved the expenditure of countless manhours in air and ground search.

I look forward to the success of your efforts to improve the safety of those who fly and the protection of those unfortunates who are exposed to accident. Crash locator beacons will save lives and will spare needless and fruitless searching.

Very truly yours,
Col. JOHN C. CAMPBELL, Jr.,
Commander.

DEPARTMENT OF TRANSPORTATION,
NATIONAL TRANSPORTATION SAFETY BOARD,
Jamaica, N.Y.
Col. JOHN C. CAMPBELL, Jr.,
New York Air National Guard,
Schenectady, N.Y.

DEAR COLONEL CAMPBELL: The following is a resume of aircraft accidents which have crashed into mountainous terrain in the immediate Albany, New York area within the last six months or have been located during this period:

Piper PA-23-250, N-5604Y, Owned and operated by New Haven Airways. Crew two commercial instrument rated pilots, High Peak Mountain (elevation 3,580 feet MSL), Tannerville, New York. Departed Albany County Airport 1414 EDT, June 13, 1967, destination Ulster County Airport, Kingston, New York. Albany weather—1500 broken, visibility 6 miles, haze, Poughkeepsie weather—estimated 900 overcast, visibility 2 miles, fog. Flight contacted Albany approach control, reporting over Hudson River and returning to Albany. The wreckage was located on June 19, 1967.

Beech 35, CF-FAS, Owned and operated by Guy Boisvert, Canadian Private Pilot, 3 passengers, Shokan Mountain (elevation 3,700 feet MSL), Margretville, New York, Departed Cartierville, Quebec, 1013 EDT, June 18, 1967. VFR flight plan to Philadelphia, Pennsylvania. Weather in general area reported as thunderstorms, low ceilings and rain. Poughkeepsie weather: (aircraft reports): 1000 broken, 1800 overcast, visibility 3 miles, light thunderstorm, haze. The wreckage was located on July 4, 1967.

Piper PA-28, N-540W, Owned by Flying West Flying Club, Crew—Private pilot and one passenger, Departed Baltimore, Maryland 1120 EDT, 6-25-67, Destination Montreal, Quebec, Albany weather: 2500 overcast, visibility 7 miles, light rainshowers. Aircraft contacted Albany approach control at 1425 EDT and reported stabilizer and gyro difficulty in precipitation east-southeast of Albany Airport. Aircraft not located to date.

Piper PA-28-150, Operator Captain (USAF), Owner Donald R. Wilkinson, New Windsor, New York, Pilot-in-command commercial, airplane single, multi-engine land, instrument, 3 passengers. Departed Orange County Airport, Walden, New York, 0835 EDT, July 1, 1967, VFR flight plan to Syracuse, New York. Weather in general area: low ceiling and fog. Wreckage located July 4, 1967, Mt. Cragmoore (elevation 3400 feet MSL), Ellenville, New York.

Cessna 172, N-9809T, Owned by Skyhaven, Inc., Rochester, New Hampshire, Operated by Private Pilot, One passenger, Departed Rochester, New Hampshire, 1445 EDT, 9-13-65, Destination Windsor Locks, Connecticut. Briefed on weather (telephone) prior to departure, weather reported as VFR marginal. The flight contacted Gardner radio. Wreckage located on July 4, 1967, on Haystack Mt. (elevation 3200 feet MSL), North Bennington, Vermont.

If further information is needed, please feel free to contact this office. The telephone number is Area Code 212, 995-3716, which is on the FTS system. It is requested that the aforegoing information not be released for public information.

The reports, when completed, can be obtained by writing to:

Accident Inquiry Section, SB-84C, National Transportation Safety Board, DOT, Bureau of Aviation Safety, 1825 Connecticut Avenue NW., Washington, D.C.

The date, location, aircraft registration number should be included with the request.

Sincerely yours,
FRANCIS MOSHER, Jr.,
Air Safety Investigator,
New York Office.

Mr. DOMINICK. Mr. President, each of these accidents resulted in a great deal of time and expense to the local civil air patrol and the local pilots who joined in the searches. Most of the time an expense would have been unnecessary with the aid of a crash-locator beacons, which the FAA reports has been perfected. So despite this fact, the FAA has not recommended regulatory action to require their use.

These reports, along with the discovery of the plane wreckage in California, must surely provide the clearest and most tragic evidence of the urgent need for FAA action to require crash-locator beacons. In most cases of lost aircraft, no record is left of the agony and suffering of the victims, but the picture is made painfully clear by reading of the 7-week ordeal of Mr. and Mrs. Olen and their daughter. Carla noted in one entry of her diary:

I hope you are happy. Search and Rescue. You haven't found us yet.

The fault lay not, however, with the searching mission, but quite simply with the fact that they did not have a crash-locator beacon. It is indeed a sad commentary on our system when three persons—and Heaven only knows how many more—suffer through such an ordeal, when our technology has provided us with an effective and available means of averting just such a disaster. How many more lives will have to be lost, Mr. President, before the FAA wakes up to its responsibility?

At this point, I ask unanimous consent to have printed in the RECORD the article entitled "Last Days of Three Crash Victims," published in the Washington Daily News of October 3, 1967, and the article entitled "Diary Records Losing Fight for Life by Girl, Mother in Crashed Plane," published in the Washington Post of October 3, 1967.

There being no objection, the articles were ordered to be printed in the RECORD, as follows:

[From the Washington (D.C.) Daily News, Oct, 3, 1967]

DIARY OF 16-YEAR-OLD GIRL: LAST DAYS OF THREE CRASH VICTIMS

REDDING, CALIF., October 3.—A girl's diary found at the scene of a plane crash told yesterday of a two-month struggle to survive in the snowy Trinity mountains which ended in death.

The bones of a man and two women were discovered Sunday by a deer hunter near the summit of 3213-foot-high Buckhorn mountain 35 miles west of here in northern California.

At least one of them had survived seven weeks in the rugged snow country after a forced landing March 11 on a flight from Portland, Ore., to Red Bluff, Calif.

The pilot was Al F. Olen, owner of the Clifford Hotel in Portland. Others were his wife, Phyllis, and her daughter, Carla Corbus, 16.

"On leg of journey to Red Bluff," said the diary's first entry. "Plane on left side in snowbank, 1215 p.m. Fuselage broken. Door ajar. Windows on right side were broken as well as windshield."

In another entry, Carla said Olen suffered a broken right arm, crushed right ribs and had pain in his back. Mrs. Olen, she said, was delirious for one day and suffered a broken right hand and left ankle.

Carla said her own injuries were an injured back, a sore right ankle and a cut on the left knee.

On April 30, she noted her 16th birthday: "I want to be rescued today," she wrote. The final entry was dated May 4 and read, "we are completely soaked."

Most of the diary, written in a flight guide book, noted weather conditions. She said the three melted snow for water but made no mention of obtaining food.

The diary also told of planes flying overhead and at one point said "Al" went for help—but apparently he returned when he failed. No mention was made of the deaths of Mr. and Mrs. Olen.

The plane was found by Floyd Bolling, a Shasta, Calif., mill worker, and his wife.

[From the Washington (D.C.) Post, Oct. 3, 1967]

DIARY RECORDS LOSING FIGHT FOR LIFE BY GIRL, MOTHER IN CRASHED PLANE

REDDING, CALIF., October 2.—A 16-year-old girl and her mother fought for two months a losing battle for their lives in the freezing carcass of their plane after it crashed on a snowy mountain peak in March, a makeshift diary disclosed tonight.

"Today is my 16th birthday. I wanted to be rescued today," the girl scribbled in the back of an airman's guide 50 days after the crash. "I hope you are happy, Search and Rescue. You haven't found us yet."

That was on April 30, and rescue never came for Carla Corbus or for her mother, Phyllis. Their scattered bones were found near the plane today. Many animal tracks were nearby.

Search parties found no trace of her stepfather, Alvin Olen, a 59-year-old hotel owner from Portland, Ore., who left the plane six days after it crashed on March 11 to seek help.

A deer hunter discovered the wreckage of the single-engine Cessna 191 about 4,000 feet up the summit of Bully Choot Mountain, about 35 miles west of Redding and a mile from a seldom-used road. Search parties had been concentrating their hunt for the missing plane miles north of the site where it actually crashed.

The three victims were on their way to San Francisco to visit Olen's oldest son, Alvin Jr., 32. The son, an airline pilot, spent 107 days in Northern California after the crash. But, he recalled here Monday, during most of that time the weather was so foul he could not fly to hunt for the lost plane.

(In Portland, the Civil Aeronautics Authority said Olen was the object of a search over Montana in 1956 on a flight from Minneapolis to Portland after he changed plans enroute. In 1954 he was fined for reckless flying after taking off from and landing in Vancouver, B.C., only by the light of automobile headlights.)

"On leg of journey to Red Bluff (Calif.)" said the diary's first entry. "Plane on left side in snowbank, 12:15 p.m. Fuselage broken. Door ajar. Windows on right side were broken as well as windshield."

The diary, with entries by both mother and daughter, then described the injuries suffered in the crash.

"Al" suffered a cut on the chin, three cuts on the forehead, a broken arm, crushed ribs on the right side, and "pain in his vertebrae."

Phyllis was "delirious one day," suffering a broken left arm, cuts on the right hand, a broken left ankle and many cuts and bruises, plus frostbitten feet.

Carla was less seriously hurt. The diary noted she "hurt in the back near her left kidney. Sore right ankle. Cut on her left knee."

Three days after the crash, Carla tried to walk through the snow but was turned back "because her feet were frozen and she had lost her shoes."

Six days after the crash, the diary said Olen left the crash scene at the 5000-foot level of the mountain.

"Al shouted ok. He crossed the gully. He was on his way for help."

A week later, an entry written by Mrs. Olen said: "Fear Al did not make it for help. Getting weak."

The diary said they melted snow for water but made only one reference to food—the day Olen left. It said simply: One glass of jelly left."

Mr. DOMINICK. Mr. President, in addition to those whose planes have crashed, we have also done some investigation to find out how many of the people who have been looking for them have also lost their lives.

In 1965, the Civil Air Patrol had three accidents with one life lost. They had test missions, practicing the procedures for rescue, in which they had five accidents and two lives lost. In 1966, they had two accidents in the CAP, although, fortunately, they did not lose any lives. In 1967, up to date, there have been two accidents, with three lives lost.

These statistics are from information given to us by Lt. Col. Charles McDonell, U.S. Air Force liaison officer for the Civil Air Patrol.

Fortunately, Mr. President, there have been no U.S. Air Force aircraft lost this year, although there has been one accident with a helicopter. But this report just points out the cost in lives, money, and time which are expended in searching for downed craft.

In the report from Colonel Campbell— which I shall not read in its entirety—it is stated that just this year, in New York, a Piper crashed on June 13, which was not located until July 19. A Beech which crashed on June 18 was not located until July 4. Another Piper which crashed on June 25 has not been located to date.

The PRESIDING OFFICER. The Senator's time has expired.

Mr. MANSFIELD. Mr. President, I ask unanimous consent that the Senator from Colorado may have 5 additional minutes.

The PRESIDING OFFICER. Without objection, it is so ordered.

Mr. DOMINICK. That aircraft has not been located to date, though it crashed on June 25.

Another Piper which went down on July 1, was found on the 4th. Also, Mr. President, there is a report of a Cessna 172 which went down on September 13, 1965, in New York, and was finally located on July 4, 1967—almost 2 years after it crashed.

Yet the FAA has stated, in reports to me, that the crash-locator beacon is not required because the pilots object to the added expense. The only way we will ever get these devices into the planes is to have a regulation that the manufacturers must incorporate them as original equipment when they build the airplane, with the cost included as part of the purchase price. For the small additional cost each pilot would have to assume in order to take care of the expense of this item, we could possibly save as much as $59 million in tax funds expended for search-and-rescue mission. It seems ridiculous to me to have the FAA time after time, state that they are not going to do anything about crash beacons because they have heard some objections from some of the aviation people, who do not wish to pay that much money. I certainly must say this makes little sense from the taxpayer's point of view, from the point of view of human suffering, or from the point of view of the country at large, in our effort to try to use updated technology to provide aviation safety at its best.

Congressional Record—House, March 12, 1969; pages 6113 through 6114

CRASH LOCATOR BEACONS

(Mr. OTTINGER asked and was given permission to extend his remarks at this point in the RECORD.)

Mr. OTTINGER. Mr. Speaker, I am introducing today, in behalf of myself and 27 colleagues, a bill directing the Federal Aviation Administration to require crash locator beacons on all civil aircraft in the United States, and also requiring all commercial aircraft, including air taxis, to be equipped with a receiving device capable of maintaining a continuous listening watch on the aeronautical international emergency frequency.

The purpose of this legislation is to end the FAA's 10-year delay in requiring the use of a device whose lifesaving capability is no longer open to question. The FAA needs no additional legislative authority to require crash locator beacons as a condition of aircraft certification. What it needs is the courage to resist pressure from certain segments of the aviation industry and I am convinced that a mandate from the Congress is necessary in that regard.

Last year, the bodies of an Oregon family whose light plane crashed in the mountains of northern California were discovered. The mother and daughter survived the crash and lived for 54 days. The father went for help and disappeared. Search planes covered the area, some flying directly overhead, but poor visibility prevented them from sighting the downed plane. The daughter's diary described in poignant detail their anguish at hearing help so close, and yet so far. A crash locator beacon would have enabled search planes to pinpoint the site of the downed aircraft for search parties on the ground. Three lives would have been saved.

A similar incident several years ago took the life of NASA flight surgeon Dr. Randolph Lovelace, whose air taxi crashed in Colorado, not far from a well-traveled highway. Dr. Lovelace survived the crash but died of exposure 5 days later. Again, a crash locator beacon would have saved a life.

Our Armed Forces in Vietnam are using crash locators with great success. The lives of more than a thousand flight crews have been saved by this little device, weighing about 2 pounds and costing under $300.

Brig. Gen. Allison C. Brooks, Commander of the Air Force Aerospace Rescue and Recovery Service, told a House subcommittee in 1967 that the ARRS in Orlando, Fla., conducted 254 search and rescue missions in 1964 involving civilian aircraft. A total of 801 people were on the missing planes. Fifty-eight were never located and six people lost their lives in the search efforts. We are currently spending about $60 million each year on air search and rescue missions and much of this is due to the amount of time involved in locating the site of a downed plane.

Recently, the FAA participated in successful tests of crash locator beacons in the Grand Canyon of Arizona and in the mountains of northern California. Two FAA DC-3's with press representatives aboard took less than 20 minutes to pinpoint a crash locator beacon planted in the Klamath Mountains about 3 miles from the spot where the Oien family perished.

A wide variety of beacons are available from at least 11 different manufacturers. They range in price from $80 to $750. Many are built to be automatically activated upon impact. Some put out both a powerful radio signal and a strobe light signal. Some have flotation characteristics, which make them particularly useful for flights over water. They are capable of sending continuous signals for 12 hours up to 30 days.

It is clear, Mr. Speaker, that crash locator beacons have been proven effec-

tive. It is clear that they are available at prices well within the means of all aircraft owners and pilots. It is clear that their mandatory use is required in view of the number of people exposed to crashes and the resources we are devoting to search and rescue missions.

By requiring all aircraft to carry crash locator beacons as a condition of certification by the FAA we are giving all pilots and passengers a meaningful measure of security. And by requiring commercial aircraft to be equipped with receivers capable of picking up the signal from a crash locator beacon, we are greatly improving the chances of speedy rescue. With lives in the balance, we can do no less than this.

Joining me today in introducing this legislation are six distinguished colleagues, who serve with me on the Committee on Interstate and Foreign Commerce, which, I hope, will act expeditiously and favorably on our bill. They are Mr. ADAMS, Mr. Moss, Mr. DINGELL, Mr. MURPHY of New York, Mr. TIERNAN, and Mr. CARTER.

Also sponsoring the legislation are Mr. ADDABBO, Mr. JACOBS, Mr. GRAY, Mr. CHARLES WILSON, Mr. GEORGE BROWN, Mr. PODELL, Mr. NIX, Mr. SISK, Mr. HELSTOSKI, Mr. ROSENTHAL, Mr. MATSUNAGA, Mr. MOORHEAD, Mr. HATHAWAY, Mr. MIKVA, Mr. PHILLIP BURTON, Mr. HALPERN, Mr. DONOHUE, Mr. CLARENCE MILLER, Mr. CLEVELAND, Mr. ROYBAL, and Mr. GLENN ANDERSON.

Congressional Record—Senate, February 26, 1970; pages 5058 through 5064

AIRPORT AND AIRWAYS DEVELOPMENT ACT OF 1969

The Senate continued with the consideration of the bill (H.R. 14465) to provide for the expansion and improvement of the Nation's airport and airway system, for the imposition of airport and airway user charges, and for other purposes.

. . .

AMENDMENT NO. 521

Mr. DOMINICK. Mr. President, I call up amendment No. 521.

The PRESIDING OFFICER. The amendment will be stated.

The assistant legislative clerk proceeded to state the amendment.

Mr. DOMINICK. Mr. President, I ask unanimous consent that further reading of the amendment be dispensed with.

The PRESIDING OFFICER. Without objection, it is so ordered; and, without objection, the amendment will be printed in the RECORD.

The amendment ordered to be printed in the RECORD, reads as follows:

After line 3, page 143, add the following new section:

"That section 601 of the Federal Aviation Act of 1958 is amended by inserting at the end thereof a new subsection as follows:

" 'Downed Aircraft Rescue Transmitters

" '(d) Minimum standards pursuant to this section shall include a requirement that downed aircraft rescue transmitters shall be installed—

" '(1) on any aircraft for use in air commerce, the manufacture of which is completed, or which is imported into the United States, after six months following the date of enactment of this subsection;

" '(2) on any aircraft used in air transportation after two years following such date; and

" '(3) on any aircraft used in air commerce after five years following such date.' "

Mr. DOMINICK. Mr. President, the printed amendment is on every Senator's desk.

I yield myself 10 minutes to explain the amendment. It may take a little more time. I do not intend to take very long. I hope that the manager of the bill will accept the amendment. If he is not going to accept it and can so indicate now, we might as well get the yeas and nays.

Mr. CANNON. Mr. President, I do not know whether I will accept the amendment. I want to hear what the Senator has to say.

Mr. DOMINICK. Mr. President, the language has been put into amendment form. It was an original bill introduced by the Senator from Washington (Mr. MAGNUSON) and me several years ago and again in the beginning of this Congress. It has not had any hearings. However, despite the fact that it has not had any hearings, I think that the evidence of the need for this is perfectly clear.

For years now, we have had general aviation aircraft go down either for mechanical reasons or because of weather or pilot error or whatever other reason it might have been.

Immediately upon that happening, and when it is discovered that they have not arrived where they intended to go, search and rescue efforts are then started. Then someone has to find out where they are. And they have continued these efforts and have spent many flying hours in doing so. They have lost people in the process of air rescue efforts. It has happened all over the country.

The cost in terms of money to the taxpayers for the Air Force and the Civil Air Patrol and the cost in terms of how many lives of people who have not been found has been absolutely extraordinary.

I think in order to put the matter in perspective, I ought to give some figures.

Starting in 1961, when inadequate records were being kept, two airplanes were reported down. Both of them were in California, or one might have been in California or Oregon. Four persons were on board. They have never been found, neither the airplanes nor the people.

In 1962, when further effort was made along this line in the way of keeping records, 11 aircraft were reported down. They have never been found. There were 16 persons on board.

In 1963, five aircraft were reported down. There were 10 people missing.

In 1964, four airplanes and five people were involved.

In 1965, 13 airplanes and 22 people were involved.

In 1966, 13 airplanes and 20 people were involved.

In 1967, 12 airplanes and 23 people were involved.

The most information I have got for 1968 is that 18 aircraft and 38 people were involved.

We do not have the figures for 1969.

I think we can see the problem this creates not only in terms of rescue efforts involved in going in to try to find these airplanes, but also the cost in human misery. Every family of each person who has been reported down simply finds that it is in a position, legally speaking, where it has a missing relative of one form or another.

In many States, the estate is tied up for over 7 years because there is no presumption of death until the 7-year period has gone by. They cannot do anything about the estate or about the property situation.

In the meanwhile, they do not know where the missing persons are, whether they are injured or dead, or whether they have simply disappeared for reasons of their own.

From 1961 to 1968, there have been a total of 78 aircraft which have totally disappeared with 139 people on board, despite all the rescue efforts that have been made.

What expense is involved? What does this mean in terms of people? I do not have the figures here immediately. However, I have put them in the RECORD before. Reciting from memory only, from 1961 to 1965 the cost to the general taxpayer in terms of the cost of operating the Aerospace Rescue and Recovery Service was $59 million.

These are just the search and rescue efforts that have been made that we know of. And in many cases the Civil Air Patrol has voluntarily carried the whole load and not even turned in the cost of their gasoline to the Federal Government.

I have some news items here which I think are pretty interesting.

Here is one of November 14, 1969. It is entitled, "It Was Terrible; Horror of 5 Crash-Stranded Days Told." The article was from Nevada City, Calif. It describes the people who were talking from hospital beds where this woman and her husband were 5 days in the aircraft waiting for someone to come and rescue them.

Here is another article entitled, "Colorado CAP Wing Halts Search for Light Plane." It tells of a missing light plane reported down between Denver and Grand Junction. It does not say how long this search went on.

I have another article entitled, "CAP's Search for Airplane is Continuing." This refers to the airplane being down between Denver and Grand Junction.

I have another article entitled, "Two Weeks in Plane Wreckage; Error in Search Saves Two." The people had been stranded for 2 weeks.

These are all 1969 clippings that I have kept. I have here an editorial from one of the papers entitled, "Protect Pilots From Themselves."

I have another clipping entitled, "Area Men Object of CAP Hunt."

Mr. President, I ask unanimous consent that the articles and editorial to which I have referred be printed at this point in the RECORD.

There being no objection, the material was ordered to be printed in the RECORD, as follows:

[From the Washington Daily News, Nov. 14, 1969]

HORROR OF 5 CRASH-STRANDED DAYS TOLD

NEVADA CITY, CALIF., November 14.—"Marvin was very strong, he handled the controls and pulled us out, but we went down again . . . lower and lower. I prayed to God to save us."

Anita Miler, 23, spoke softly from a hospital bed. A few feet away her husband Marvin, 25, mumbled thru the wires binding his broken jaw: "It wasn't the plane's fault."

The Vancouver, Wash., couple, en route from Reno, Nev., to Disneyland near Anaheim, Calif., crashed last Friday on a mountainside and survived five days on melted snow and dried soup.

They were rescued after Mr. Miler struggled eight miles thru foot-deep snow with a broken ankle, jaw and wrist to a mountain resort.

"We crashed and I looked up and here I was and I was all right, and I turned to Marvin and I said, 'Honey, let's get out of here,' she said.

But her husband was unconscious.

"I looked around and there a few feet away was a cabin. It took me a long time to get out of the plane; I was all pinned in," Mrs. Miler said.

"I came back for Marvin. He was out, he couldn't hear what I said. I helped him into the cabin.

"He scared me so because the blood was just running out of his ear. It was terrible. He just kept saying, 'What happened?'

BUILT FIRE

"I helped him into the cabin. When we got in, there was a stove. I pulled paper from the wall. I had some matches. I pulled out the cupboard and shelves and burned every piece of wood I could."

Mr. Miler was delirious for a day, while his wife melted snow in a soft drink can and prepared dried soup. On the second day, he recovered.

For three days the couple stayed close to the cabin. They burned an abandoned building at one point to attract rescuers but nobody noticed, despite the fact one plane came so close to the crash site "we could have hit it with a rock."

On Wednesday, Mr. Miler set out to seek help and was found wandering along the roadside about two miles from Sierra City, a mountain village.

A sheriff's rescue vehicle then went in to bring out Mrs. Miler.

COLORADO CAP WING HALTS SEARCH FOR LIGHT PLANE

The Colorado Wing of the Civil Air Patrol (CAP) Sunday night called off its search for a missing light plane piloted by a Grand Junction, Colo., man because of a lack of leads.

The plane, a Cessna 150 flown by Glenn Scott, 69, vanished Oct. 31 on a flight from Denver to Grand Junction. Capt. Harlan Cook, CAP information officer, said Scott had 100 hours of flying experience.

Cook said the search will be reopened if new leads are found.

During the first weekend of the hunt, planes were kept on the ground by bad weather. But fair weather made the search a full-scale effort every day last week, with as many as 15 planes and 25 ground parties participating each day.

The planes systematically covered a wide area along the entire probable flight path of the missing plane. Cook said the CAP's effort was hampered by new snow, which totaled 19 inches in much of the search area.

CAP'S SEARCH FOR AIRPLANE IS CONTINUING

The Colorado Civil Air Patrol continued Saturday its search for a small aircraft believed down between Denver and Grand Junction.

The green Cessna 150, piloted by Glen Scott of Grand Junction and bearing the number 50938, left Denver for Grand Junction at 10:15 a.m. Friday. The aircraft carried 3½ hours of fuel for the 2½-hour flight, the CAP said.

The air search, headed by mission co-ordinator Maj. Gene Wirth, will resume when weather permits. Meanwhile, ground parties are continuing their search in the Winter Park area.

TWO WEEKS IN PLANE WRECKAGE; ERROR IN SEARCH SAVES TWO

JACKSON, CALIF.—Two men who spent two weeks in the wreckage of a light plane with the body of the pilot are safe today because of an erroneous smoke report and the determination of friends.

Neither Eugene Ebell, 33, nor Robert Staar, 17, suffered major injury from the Jan. 11 crash or their 15 days without food. Pilot Donald Shaver was killed in the crash in the Sierra Nevada mountains.

Ebell had chartered the plane in Fresno, the hometown of all three men, to fly to Elko, Nev., to pick up the body of an uncle who was to be buried in Fresno. Staar, a friend of the pilot, went along for the ride.

Ebell said the pilot tried to turn back over the Sierra Nevada because the plane's wings were icing but, in turning, the plane lost too much altitude and crashed.

Ebell and Staar were rescued by helicopter yesterday from a rugged canyon 35 miles east of Jackson after Staar was spotted from the air.

They said they had heard and seen search planes regularly, but none came far enough up the mountains to see them. The crash site was near the 7,000-foot level of the Sierra about 180 miles east of San Francisco.

Staar set out Sunday to get help.

At the same time, searchers shifted their aerial hunt to the east because of an apparently erroneous report of smoke. On their way to the area yesterday, Doyle Hawkins and helicopter pilot George Wurzburg spotted Staar beside a log where he had slept overnight after walking 3½ miles.

As many as 20 planes a day had searched the Sierra for the wreckage the first week, then gave up. Friends and relatives of the missing men collected $1,400 and hired the helicopter last Friday to continue the aerial hunt.

Doctors at Amador Hospital said the survivors were treated for exposure and minor frostbite. Ebell also had some crushed ribs.

PROTECT PILOTS FROM THEMSELVES

The white vastness of Corona Pass stretched onward for miles beneath search planes Sunday that pored over its bleakness in search of a small private plane that ended its last flight Friday with a deathly plunge into a mountainside.

Finally, after hours of looking, a plane spotted a clump of darkness in the snow.

A few hours later, ground crews pulled the bodies of a California couple from the wreckage.

The plane apparently crashed shortly after takeoff from Stapleton Field in Denver at 10:16 a.m. Friday.

Yet searchers were faced with the tremendous task of combing hundreds of square miles encompassing the flight pattern filed by the plane's pilot.

This time, there were no survivors. But there have been other times when there were. And there will be others.

Current legislation proposed by U.S. Sen. Peter Dominick and State Rep. Ted Bryant can eliminate the ever present danger of persons surviving a crash only to die of exposure or lack of medical aid.

Mandatory installation of crash locater beacons, small battery powered pieces of equipment that shoot out a life saving beam, would end the hours, days and months of waiting for help that have cost many their lives.

The pilot of the plane that crashed Friday had at least filed a flight pattern that led searchers to the crash site in a relatively short time. Others have never been found.

But, had a functioning crash locater beacon been aboard, the crash could have been found in a matter of short hours. And any survivors could have been rescued.

Despite the apparent need for required rescue equipment, there looms a bigger, more complex issue that could be combined with material equipment not only to save lives after crashes, but to prevent crashes.

Colorado's mountains have for years claimed the lives of pilots who have had too little, or no experience, in traversing them.

The intricacies of mountain flying, particularly in single engine planes, is too apparent to the state residents who read almost weekly of another pilot who "thought he could make it."

One does not receive a second chance when attempting to climb over 12,000 foot peaks while being pulled from below by unpredictable down drafts.

So the essence of air safety points in more than one direction. It is time meaningful legislation began to probe effectively all the possibilities.

And a pertinent direction should be that of specialized training for persons who attempt to navigate the Rocky Mountains from the air. Without this special training, death from the skies will continue.

AREA MEN OBJECT OF CAP HUNT

Members of the Colorado Wing of the Civil Air Patrol (CAP) Friday joined a three-state search for a missing plane carrying three Denver-area men on a flight from Denver to El Paso, Tex.

The men, all Martin Co. engineers, were identified as Ted R. Jones, 35, of 6591 S. Marion St., Arapahoe County, the pilot; Eugene W. Harker, 38, of 5065 Juniper St., Bow Mar, and William DeVos, 43, of 3453 W. Bowles Ave., Littleton.

Capt. Harlan Cook, Colorado CAP information officer, said they took off from Stapleton International Airport at 11:15 p.m. Wednesday and were to arrive in El Paso at 2:15 a.m. Thursday.

He said no report of the plane, a two-engine Cessna 310, had been received since take-off. Cook said it was reported that they were going on a fishing trip in the vicinity of Navarro, N.M.

Cook said 12 CAP aircraft and three ground parties began the Colorado phase of the hunt early Friday on a full-scale basis. Colorado units did their first searching Thursday afternoon, along with CAP personnel in New Mexico and Texas.

Cook said the Colorado searchers Friday

were concentrating on the probable flight course between Stapleton and Pueblo, Colo. The missing plane was to have passed over or near Pueblo, and Las Vegas and Corona, N.M.

Federal Aviation Administration officials in Denver said severe icing conditions existed on the missing plane's flight course at the time it was in the air.

Mr. DOMINICK. What am I trying to do by this amendment? I am trying to say that original general aviation aircraft when they are manufactured must have on them a locator beacon. When they go down, it automatically emits a signal. And anyone tuned in on this signal, which is either 121.5 or 243.0, can home in on the transmitter and find it within a matter of minutes. To give an example of whether or not it works, we had a test outside of Aspen, Colo., at a time when I happened to be flying my own airplane. I was notified a test was going to be made. I did not have a homing beacon of that frequency I could use. I had a general idea where I was going to be, somewhere near this Aspen, Colo., mountainous terrain. I tuned in and by simply using the volume control on my receiver, using this signal, within 15 minutes I was within a quarter of a mile from where it was and I did not have a homing beacon. The method simply is that when the search plane goes away from it, it disappears and when the search plane comes toward it, it increases in volume and so you can locate where the particular instrument is.

The objections we have had to this particular proposal largely have been from those people who say this type requirement should not be mandatory, that it should be voluntary. The difficulty with that is that all pilots, including myself, are basically optimistic. One has to be optimistic if he is in politics, if he is a flyer, or if he is in the mining game; otherwise no one would go into them. One has to figure he is going to win. This is especially true in being a pilot. So they have not put in this equipment.

There have been proposals by the FAA that they be required over areas such as the desert or large bodies of water. If that method is going to be followed the difficulty is there would have to be an army of inspectors to enforce it. In addition, there would be great difficulty in trying to find out where they could be picked up and returned again; whether it is going to be possible to orient the rental instrument—in other words, whether they are in proper working condition when they are rented.

The estimated cost at the present time of installing these instruments as new equipment in aircraft is between $250 and $300 per airplane. If someone is buying a new airplane, and they are sold

every day, the cost of $250 to $350 could be relatively easily absorbed, in my judgment, by the manufacturer so it would not be very much of a cost increase; and if it is not absorbed, in terms of lifesaving devices it is not going to be the difference in whether a pilot buys the airplane or not. All one would have to do is go down once and have this signal work and he will know how important this signal is to anyone in the airplane or to the families they have left at home.

Mr. President, I have just a couple of other points that I wish to make and then I shall reserve the remainder of my time.

The PRESIDING OFFICER. The time of the Senator has expired.

Mr. DOMINICK. Mr. President, I yield myself 4 additional minutes.

The PRESIDING OFFICER. The Senator from Colorado is recognized for 4 additional minutes.

Mr. MAGNUSON. Mr. President, will the Senator yield at that point?

Mr. DOMINICK. I am happy to yield to the Senator from Washington.

Mr. MAGNUSON. Mr. President, as the Senator knows, I have been interested in this matter. The amount the Senator mentioned is the present going price. However, we had some testimony to the effect that if this was going to be more widely used then they would be able to bring the cost down and as they have more orders of this type, the manufacturers, whoever they may be, would be able to produce these much, much cheaper.

Mr. DOMINICK. That is correct.

Mr. MAGNUSON. They would be installed as standard equipment.

Mr. DOMINICK. Yes. As a matter of fact, I have had information from some people who have been in to see me on this matter because I have been very active on it, and they hope to get it down to $50.

Mr. MAGNUSON. The testimony we had was to the same effect.

Mr. GOLDWATER. Mr. President, will the Senator yield?

Mr. DOMINICK. I yield.

Mr. GOLDWATER. Mr. President, to sharpen up the focus on this point, I recall that when transponders first became available for private aircraft, the price was $3,500. Now that they are becoming mandatory and can be more or less mass produced, they are being offered for under $1,000.

Knowing something about the electronics involved in a locator beacon such as this, I feel certain that when they are required to be placed on aircraft, they could be procured for between $50 and $100.

Mr. President, I commend the Senator for introducing this amendment. I know there is opposition to it, but living in the Rocky Mountain region and having participated in many searches for aircraft and having lost good friends in lost aircraft, I think it is an important measure.

I might ask the Senator if it is true that considering just the great many hours that the Civil Air Patrol has spent on searches and if we assume the ridiculously low price—I would say almost impossible price—of $10 an hour, we are talking about something close to $2 million in just the cost of gas that has been spent on these searches. Am I correct?

Mr. DOMINICK. The Senator is correct. That does not cover the cost of the Air Force when they go out and also participate in search efforts.

Mr. GOLDWATER. They do. We have a group at the Air Force base near Phoenix that goes out on all searches. The Civil Air Patrol is not the only group that goes out. We have sheriffs' air posses that participate. I do not think there is an aircraft owner in Arizona that does not have his aircraft available immediately for searches.

Probably in the Senator's State and in Wyoming, Alaska, Nevada, and the Rocky Mountain States, we lose more airplanes every week than are lost in the rest of the country in a year. This comes close to home to all of us.

Mr. DOMINICK. I appreciate the Senator's comments.

Mr. President, I might say it is very interesting. We have the number of hours for Aerospace Rescue and Recovery Service in 1968. The table is broken down among Eastern, Central, Western, and Alaskan areas. In the Eastern area there were 28 aircraft missing more than 24 hours before they were found. The Eastern area includes the States of Maine, Vermont, New Hampshire, Massachusetts, Connecticut, Rhode Island, New York, Pennsylvania, New Jersey, Maryland, West Virginia, Ohio, Kentucky, Virginia, Tennessee, North Carolina, South Carolina, Georgia, Florida, Alabama and Mississippi.

There were only 25 aircraft missing more than 24 hours in the Central Region. In the Western Region there were 33 aircraft missing more than 24 hours. The Western Region includes the States of New Mexico, Arizona, Utah, Nevada, California, Oregon, Idaho and Washington.

The PRESIDING OFFICER. The time of the Senator has expired.

Mr. DOMINICK. I yield myself 5 additional minutes.

Colorado is included in the Central area. In 1969 we had quite a number of planes that went down there. We plotted a map and put it up before the Radio

Technical Commission for Aeronautics meeting about 2 years ago when I made a talk before them in Washington. The map showed airplanes down more than 3 days. There were more of them in the area of South Carolina, Florida, and Georgia than anywhere else in the country, which I could not believe. I thought there would be more in our area or in the area of Oregon and Washington. However, I assume that is because of the lakes and marshes in the Southeast.

There may have been some of those planes that decided to take off and not tell anybody. That is always a possibility. If they had these locator beacons, we could find them.

Mr. President, I ask unanimous consent to have printed at this point in my remarks the table showing the Aerospace Rescue and Recovery Service statistics to which I have referred.

There being no objection, the table was ordered to be printed in the RECORD, as follows:

AEROSPACE RESCUE AND RECOVERY SERVICE, 1968

Eastern A.R.R.C.: (includes states of Maine, Vermont, New Hampshire, Massachusetts, Connecticut, Rhode Island, New York, Pennsylvania, New Jersey, Maryland, West Virginia, Ohio, Kentucky, Virginia, Tennessee, North Carolina, South Carolina, Georgia, Florida, Alabama and Mississippi.)

Aircraft missing more than 24 hours	28
Of which were missing more than 3 days	20
Of that total how many never been found?	5
Total search hours:	
U.S. Air Force	54
CAP	6,830
Total	6,884

Central A.R.R.C: (includes states of Michigan, Indiana, Illinois, Wisconsin, Minnesota, North Dakota, South Dakota, Montana, Iowa, Nebraska, Wyoming, Colorado, Kansas, Missouri, Arkansas, Oklahoma, Louisiana and Texas.)

Aircraft missing more than 24 hours	25
How many were missing more than 3 days?	15
How many never found?	4
Total search hours:	
U.S. Air Force	891
CAP	8,109
Total	9,000

Western A.R.R.C.: (includes states of New Mexico, Arizona, Utah, Nevada, California, Oregon, Idaho and Washington.)

Aircraft missing more than 24 hours	33
How many missing more than 3 days?	18
How many never found?	6
Total search hours:	
U.S. Air Force	1,071
CAP	7,368
Total	8,439

Alaska:

Aircraft missing more than 24 hours__	40
How many were missing more than 3 days? _____	6
How many never found?_____	3
Total search hours:	
U.S. Air Force_____	806
CAP _____	2,316
Total _____	3,122

RECAP

Search hours flown by Air Force_____	2,822
Search hours flown by Civil Air Patrol	24,623
Total search hours_____	27,445
Crashed aircraft located_____	260
Downed aircraft never found_____	18

Mr. ALLOTT. Mr. President, will the Senator yield?

Mr. DOMINICK. I yield.

Mr. ALLOTT. Mr. President, I thank my distinguished colleague for yielding, because I have followed his interest in this amendment for a long time, and I wholeheartedly support it. I think the editorial to the effect that pilots must be protected from themselves brings up the main issue that is involved.

I would like to ask the Senator one question with respect to the cost of this proposal. The Senator from Arizona has mentioned the rescue efforts of the Air Force and the National Guard. We have spoken of the CAP. In addition to these efforts, I recall from my experience with flying that almost every private airplane that was on any small airport anywhere near a downed aircraft would join in the search for the airplane.

Mr. DOMINICK. The Senator is totally correct.

Mr. ALLOTT. So there is really no way of adding up the total amount that is spent for the search and rescue efforts. We have special problems in the Rocky Mountain region with those who have not had any experience with the unique flying conditions which exist there. Some experienced flyers have gone down in these mountains, because they were unfamiliar with the updrafts and downdrafts peculiar to Rocky Mountain flying.

When there is such a locator facility available within the cost parameters talked about here, it seems outrageous to spend all this money in search-and-rescue operations, when the lost plane could have been found if a marker beacon had been used on the plane.

Mr. DOMINICK. I certainly want to thank my colleague for bringing these points up, because they dramatize the problems we have. We have not even talked about the ground searches that go on in a great many areas. Someone says he heard a low-flying airplane in

bad weather, and the airplane does not show up. As a result, there are ground searches made.

The PRESIDING OFFICER. The time of the Senator has expired.

Mr. DOMINICK. I yield myself 5 additional minutes.

In order to be totally fair, one of the problems we have had to date with this particular system is the question of who will be listening. It is all right to put up a signal, but the question is, who is going to be listening? The interesting thing is that the FAA is in the process of completely equipping its own aircraft so they can home in.

The other procedure that can be followed is to include it as a part of the NASA satellite concept. This has not been done yet because of the budget problems we have, but with a satellite overhead which could pick these programs up, within an hour the signals of the aircraft that was down could be pinpointed. It is really a quite extraordinary development.

I hold in my hand an article written by Dan Partner, a very able reporter for the Denver Post, written on October 19, 1969, in which he mentioned the possibility of using the satellite system for air traffic control. It can also be used for the air rescue effort.

I ask unanimous consent to have the article printed in the RECORD.

There being no objection, the article was ordered to be printed in the RECORD, as follows:

TRAFFIC CONTROL SATELLITES URGED
(By Dan Partner)

A satellite system for use in air traffic control is emerging as a practical application of space technology that has produced the communications and weather satellite programs.

Given a high priority, a system of satellites could be orbiting the earth by 1976 that could pinpoint the positions of thousands of aircraft expected to be clogging the domestic and international airlanes. The system would be similar, but considerably more advanced, than the four-satellite system now in use for surface ships.

TRW Inc. engineers are working on a plan that would permit Federal Aviation Administration air traffic control centers to determine positions of aircraft to an accuracy of 50 feet through data flashed from space at one-second intervals. Technology for the system is available, says David D. Otten, advanced systems manager for control and navigation satellites for TRW's Systems Group.

A small antenna and transmitter for use aboard aircraft would cost $400 and weigh three pounds. In addition, the satellite system would provide precise radio navigation to planes at a cost of about $5,000 each, Otten estimates.

The system would require six satellites to cover the United States and from 12 to 15 to service worldwide air traffic routes. Otten estimates the cost at from $54 million and $66 million, including development, hardware, launch and operational expenditures. Each satellite would have a lifetime of about five years.

Meanwhile, the FAA and the Department of Transportation are beating the drums for passage of the aviation facilities expansion act, now before Congress. In an article, "Logjams in the Sky," in the September issue of FAA Aviation News, Transportation Secretary John Volpe wrote:

Passengers carried in 1968 by U.S. airlines amounted to 75 per cent of the nation's population. At the rate passenger traffic is increasing, the number of passengers carried will surpass the population within a short time.

In the general aviation category, private fleets are doubling every decade. This segment of aviation will represent 10 per cent of the gross national product by 1980.

Air freight hauled by commercial airlines jumped an unprecedented 21 per cent last year over the previous 12 months.

The proposed legislation, Volpe contends, maintains that if present growth in aviation is to continue, then both commercial and general aviation interests must share in the development costs to improve and update U.S. airport and airways facilities.

Says FAA Administrator John Shaffer: "The expansion of our air traffic control system has fallen far short of matching the growth in air traffic. More than two-thirds of the nation's 3,200 airports are in need of landing area improvements, and 900 more airports are going to be needed before 1980."

The administration's airport-airways program has a user tax base which would set up a designated account to protect the funds for use on the airports and airways. The bill establishes a federal commitment to a 10-year $2.5 billion grant-in-aid program. It authorizes $1.25 billion over the next five years, starting with $180 million in fiscal year 1970 and $220 million in 1971.

Mr. DOMINICK. In addition, as I pointed out before, if it is known that a flier is going from one point to another, either by his family or through a flight plan, and we get a report that the plane is down, it is not only possible, but it will inevitably happen that any private aircraft going through that area will start monitoring those signals. By the volume control one can, generally speaking, pick it up and determine where it is.

Second, it is totally feasible, and I think it is highly possible, to get the commercial airliners which are crisscrossing the country to install a little receiver—this is not required in the bill; I am just pointing out what can be done—with a pinpoint light on the dashboard. When a signal is picked up it will flicker. All the pilot has to do is report it to the nearest FAA flight service station. They in turn can start the air rescue effort. I have talked with some of the officials with respect to this matter. They do not want to go ahead with it

until other concepts can be explored, because of the cost involved. I do not blame them for it. But if we can go ahead, we will be that far ahead of the game.

Mr. CANNON. Mr. President, will the Senator yield?

Mr. DOMINICK. I yield.

Mr. CANNON. I wanted to raise a few points in colloquy with the Senator.

I am sure the Senator is aware of the fact that the FAA proposed that very thing, under its rulemaking authority, in 1968, and it heard such an uproar from the users and pilots that they did not do anything further with it. So, in effect, that has been abandoned.

It seems to me there are two or three weaknesses in that proposal, and I would like to have the Senator respond to those, if he would not mind.

Mr. DOMINICK. May I respond to the first one first?

Mr. CANNON. Yes.

Mr. DOMINICK. They proposed at that time to preposition some of these locator beacons at base operations throughout the country, where they could be picked up on a rental basis and put in the aircraft and could be used when going over a deserted area, or desert land, or a body of water. That system is not going to work. Many of the objections did not go to putting that device in; they went to the question that the system would not work and the money would be put up for nothing. I myself objected to it.

Mr. CANNON. Two points disturb me. One of them is that there are a number of small airplanes, two-seater airplanes, around the country that people plan on using no more than 50 miles away from their home. They like to fly for a little pleasure and sport. To that type of airplane you would add something that will add substantially to the cost of the airplane, without reducing any appreciable risk.

Mr. DOMINICK. May I answer that question first?

Mr. CANNON. Yes.

Mr. DOMINICK. I am not sure of the exact percentage, but I think it is right. Approximately 60 percent of the accidents that happen in general aviation occur within 20 miles of the airport. In my part of the country, and I am sure in the Senator's part of the country, I know we have cases in which a person has gone off the airport, has disappeared in a cloud, has crashed, and has not been found for a week. So it is just as important in training planes as in planes which can be used for cross-country flying, so they can be found if they go down.

Mr. CANNON. To go to the figure used by the Senator, I would go further and say that 60 percent of mishaps happen within 1 mile of the airport. A good many of them happen right at the airport. What the Senator is trying to do is to have some kind of device that would help in locating a lost airplane.

Mr. DOMINICK. That is correct.

Mr. CANNON. The Senator does not make any distinction with respect to the larger jets. Take the commercial jets. I would say that the system for tracking them and knowing where they are at any given time is much more accurate than any locator beacon system such as would be installed in the small airplanes as you suggest. It seems to me it would be a waste of corporate money of the commercial airline industry to have to put this kind of equipment in the commercial jet airplanes.

Mr. DOMINICK. Let me say that I am not anchored in on phase 3. It gives them 5 years to put it in. I am perfectly willing to take that provision out. The only reason it was put in there originally is that many aircraft used in air commerce were not jets. Particularly was this true 2 years ago, when we started developing this device.

If the Senator will feel happier about it, I am perfectly willing to modify the amendment to take out paragraph 3.

As to the propeller airplane problem, the commercial airliner going out over water, extensive water hazards, and things of that kind, I can see how perhaps we need something to deal with those matters, but since almost all of them are flying almost totally under instrument flight conditions, where they are monitored all the way, I think this may be asking a little more than we should, and I am perfectly willing to modify the amendment to that extent.

The PRESIDING OFFICER. The Senator's time has again expired.

Mr. DOMINICK. I yield myself 2 additional minutes.

Mr. President, I ask unanimous consent to eliminate from my amendment subparagraph (3) on page 2.

The PRESIDING OFFICER. The Senator has the right to modify his amendment.

Mr. DOMINICK. I so modify my amendment, and I strike the semicolon and the word "and" in line 5, and insert a period after the word "date".

The PRESIDING OFFICER. The amendment will be modified as the Senator has specified.

Mr. DOMINICK. I think the suggestion of the Senator from Nevada is reasonable, and I am happy to accept it, and have so modified my mendment.

Mr. President, I ask unanimous consent to have printed in the RECORD a table that I have showing aircraft which

have been missing from 1961 through 1967, together with the number of people on board and the States from which they were declared missing.

There being no objection, the table was ordered to be printed in the RECORD, as follows:

MISSING AIRCRAFT

Area	Date	Persons on board
California, Oregon	Dec. 1, 1961	2
California	Mar. 14, 1967	2
Total		4
Louisiana, Texas	Jan. 4, 1962	3
South Carolina	Mar. 1, 1962	1
Oregon	Mar. 17, 1962	1
Alaska	June 5, 1962	1
Michigan	June 28, 1962	1
North Carolina	July 22, 1962	4
Michigan	Aug. 18, 1962	1
Do	Sept. 1, 1962	1
Alaska	Oct. 16, 1962	1
Do	Oct. 18, 1962	1
Washington	Nov. 10, 1962	1
Total		16
Utah, Colorado	Jan. 9, 1963	4
Utah, Nevada, California	Mar. 28, 1963	2
Oregon	July 20, 1963	1
Washington	Aug. 28, 1963	1
Michigan, New York	Nov. 3, 1963	2
Total		10
Washington	Jan. 27, 1964	1
North Illinois	Feb. 15, 1964	1
South Carolina	May 3, 1964	1
Oregon, Washington	June 15, 1964	2
Total		5
Florida, Alabama	Jan. 3, 1965	4
Washington	Jan. 29, 1965	1
Do	May 17, 1965	4
Alaska	June 6, 1965	1
South Florida	July 7, 1965	1
Louisiana	July 12, 1965	1
South Carolina	Sept. 5, 1965	1
Kentucky, North Carolina	Sept. 7, 1965	1
Alaska	Sept. 13, 1965	2
West Massachusetts	Sept. 14, 1965	2
South Florida	Nov. 1, 1965	2
Do	Dec. 7, 1965	1
California	Dec. 10, 1965	2
Total		22
Maine, Vermont, New Hampshire	Mar. 20, 1966	1
Maine	Apr. 2, 1966	2
South Carolina	do	1
New York, Massachusetts	Apr. 27, 1966	2
South Carolina	May 10, 1966	1
Arizona	June 21, 1966	1
North Carolina	June 28, 1966	2
Do	July 14, 1966	1
Florida, Mississippi, Louisiana	Sept. 20, 1966	2
Alaska	Sept. 23, 1966	1
Do	Oct. 9, 1966	1
Alabama, Georgia	Nov. 8, 1966	1
Ohio	Dec. 20, 1966	4
Total		20
Florida	Jan. 15, 1967	4
Michigan	do	3
North Carolina	Apr. 24, 1967	2
South Texas, Mexico	Apr. 27, 1967	1
Utah, Nevada, California	June 3, 1967	3
Alaska	June 14, 1967	1
Florida	July 8, 1967	1
Missouri	Aug. 26, 1967	2
Florida	Oct. 8, 1967	1
South Florida	Oct. 11, 1967	2
Arkansas, Texas	Oct. 14, 1967	1
Ohio, Kentucky, Tennessee, Georgia	Dec. 23, 1967	2
Total		23

Mr. DOMINICK. I reserve the remainder of my time.

Mr. MANSFIELD. Mr. President, I suggest the absence of a quorum.

The PRESIDING OFFICER. On whose time?

Mr. CANNON. Mr. President, the time may be taken from my time.

The PRESIDING OFFICER. The clerk will call the roll.

The assistant legislative clerk proceeded to call the roll.

Mr. CANNON. Mr. President, I ask unanimous consent that the order for the quorum call be rescinded.

The PRESIDING OFFICER. Without objection, it is so ordered.

Mr. DOMINICK. Mr. President, I further modify my amendment, and send the modified amendment to the desk. I shall read it now, so that we can be sure Senators know what it is:

After line 3, page 143, add the following new section: That section 601 of the Federal Aviation Act of 1958 is amended by inserting at the end thereof a new subsection as follows:

"Downed Aircraft Rescue Transmitters

"(d) Minimum standards pursuant to this section shall include a requirement that downed aircraft rescue transmitters shall be installed—

"(1) on any aircraft for use in air commerce, except jet aircraft used in commercial transport, the manufacture of which is completed, or which is imported into the United States, after six months following the date of enactment of this subsection;"

I think, if the Senator from Nevada does not mind, we will change that to "one year" instead of "six months".

Mr. CANNON. Very well.

The PRESIDING OFFICER. The amendment will be modified as specified.

Mr. DOMINICK. So it would read:

After one year following the date of enactment of this subsection;

And then continuing:

(2) on any aircraft used in air transportation after three years following such date.

Subsection (3) would be stricken.

As such, it is my understanding that the Senator from Nevada will accept the amendment.

Mr. CANNON. Mr. President, I am willing to accept the amendment as now modified by the distinguished Senator from Colorado.

I yield back the remainder of my time.

The PRESIDING OFFICER. Does the Senator from Colorado yield back the remainder of his time?

Mr. DOMINICK. I yield back the remainder of my time.

The PRESIDING OFFICER. The remaining time having been yielded back, the question is on agreeing to the amendment (No. 521) of the Senator from Colorado (Mr. DOMINICK), as modified.

The amendment, as modified, was agreed to.

FAA Regulations for ELTs

The following are excerpts from Federal Aviation Regulations showing some examples of the resulting regulations following the Bill Congress passed in 1970.

From Title 14 of the Code of Federal Regulations, Chapter 1, Part 91, "General Operating and Flight Rules"—

§91.207 Emergency locator transmitters.

(a) Except as provided in paragraphs (e) and (f) of this section, no person may operate a U.S. registered civil airplane unless—

(1) There is attached to the airplane an approved automatic type emergency locator transmitter that is in operable condition for the following operations, except that after June 21, 1995, an emergency locator transmitter that meets the requirements of TSO-C91 may not be used for new installations:

(i) Those operations governed by the supplemental air carrier and commercial operator rules of parts 121 and 125;

(ii) Charter flights governed by the domestic and flag air carrier rules of part 121 of this chapter; and

(iii) Operations governed by part 135 of this chapter; or

(2) For operations other than those specified in paragraph (a)(1) of this section, there must be attached to the airplane an approved personal type or an approved automatic type emergency locator transmitter that is in operable condition, except that after June 21, 1995, an emergency locator transmitter that meets the requirements of TSO-C91 may not be used for new installations.

(b) Each emergency locator transmitter required by paragraph (a) of this section must be attached to the airplane in such a manner that the probability of damage to the transmitter in the event of crash impact is minimized. Fixed and deployable automatic type transmitters must be attached to the airplane as far aft as practicable.

✶ ✶ ✶ ✶ ✶

From the FAA *Aeronautical Information Manual*, Chapter 6, "Emergency Procedures"—

6–2–4 Emergency Locator Transmitter (ELT)

a. General.

1. ELTs are required for most General Aviation airplanes.

Reference: 14 CFR Section 91.207.

2. ELTs of various types were developed as a means of locating downed aircraft. These electronic, battery operated transmitters operate on one of three frequencies. These operating frequencies are 121.5 MHz, 243.0 MHz, and the newer 406 MHz. ELTs operating on 121.5 MHz and 243.0 MHz are analog devices. The newer 406 MHz ELT is a digital transmitter that can be encoded with the owner's contact information or aircraft data. The latest 406 MHz ELT models can also be encoded with the aircraft's position data which can help SAR forces locate the aircraft much more quickly

after a crash. The 406 MHz ELTs also transmits a stronger signal when activated than the older 121.5 MHz ELTs.

(a) The Federal Communications Commission (FCC) requires 406 MHz ELTs be registered with the National Oceanic and Atmospheric Administration (NOAA) as outlined in the ELT's documentation. The FAA's 406 MHz ELT Technical Standard Order (TSO) TSO-C126 also requires that each 406 MHz ELT be registered with NOAA. The reason is NOAA maintains the owner registration database for U.S. registered 406 MHz alerting devices, which includes ELTs. NOAA also operates the United States' portion of the Cospas-Sarsat satellite distress alerting system designed to detect activated ELTs and other distress alerting devices.

(b) In the event that a properly registered 406 MHz ELT activates, the Cospas-Sarsat satellite system can decode the owner's information and provide that data to the appropriate search and rescue (SAR) center. In the United States, NOAA provides the alert data to the appropriate U.S. Air Force Rescue Coordination Center (RCC) or U.S. Coast Guard Rescue Coordination Center. That RCC can then telephone or contact the owner to verify the status of the aircraft. If the aircraft is safely secured in a hangar, a costly ground or airborne search is avoided. In the case of an inadvertent 406 MHz ELT activation, the owner can deactivate the 406 MHz ELT. If the 406 MHz ELT equipped aircraft is being flown, the RCC can quickly activate a search. 406 MHz ELTs permit the Cospas-Sarsat satellite system to narrow the search area to a more confined area compared to that of a 121.5 MHz or 243.0 MHz ELT. 406 MHz ELTs also include a low-power 121.5 MHz homing transmitter to aid searchers in finding the aircraft in the terminal search phase.

(c) Each analog ELT emits a distinctive downward swept audio tone on 121.5 MHz and 243.0 MHz.

(d) If "armed" and when subject to crash-generated forces, ELTs are designed to automatically activate and continuously emit their respective signals, analog or digital. The transmitters will operate continuously for at least 48 hours over a wide temperature range. A properly installed, maintained, and functioning ELT can expedite search and rescue operations and save lives if it survives the crash and is activated.

(e) Pilots and their passengers should know how to activate the aircraft's ELT if manual activation is required. They should also be able to verify the aircraft's ELT is functioning and transmitting an alert after a crash or manual activation.

(f) Because of the large number of 121.5 MHz ELT false alerts and the lack of a quick means of verifying the actual status of an activated 121.5 MHz or 243.0 MHz analog ELT through an owner registration database, U.S. SAR forces do not respond as quickly to initial 121.5/243.0 MHz ELT alerts as the SAR forces do to 406 MHz ELT alerts. Compared to the almost instantaneous detection of a 406 MHz ELT, SAR forces' normal practice is to wait for either a confirmation of a 121.5/243.0 MHz alert by additional satellite passes or through confirmation of an overdue aircraft or similar notification. In some cases, this confirmation process can take hours. SAR forces can initiate a response to 406 MHz alerts in minutes compared to the potential delay of hours for a 121.5/243.0 MHz ELT.

3. The Cospas-Sarsat system has announced the termination of satellite monitoring and reception of the 121.5 MHz and 243.0 MHz frequencies in 2009. The Cospas-Sarsat system will continue to monitor the 406 MHz frequency. What this means for pilots is that after the termination date, those aircraft with only 121.5 MHz or 243.0 MHz ELTs onboard will have to depend upon either a nearby Air Traffic Control facility receiving the alert signal or an overflying aircraft monitoring 121.5 MHz or 243.0 MHz detecting the alert. To ensure adequate monitoring of these frequencies and timely alerts after 2009, all airborne pilots should periodically monitor these frequencies to try and detect an activated 121.5/243.0 MHz ELT.

b. Testing.

1. ELTs should be tested in accordance with the manufacturer's instructions, preferably in a shielded or screened room or specially designed test container to prevent the broadcast of signals which could trigger a false alert.

2. When this cannot be done, aircraft operational testing is authorized as follows:

(a) Analog 121.5/243 MHz ELTs should only be tested during the first 5 minutes after any hour. If operational tests must be made outside of this period, they should be coordinated with the nearest FAA Control Tower or FSS. Tests should be no longer than three audible sweeps. If the antenna is removable, a dummy load should be substituted during test procedures.

(b) Digital 406 MHz ELTs should only be tested in accordance with the unit's manufacturer's instructions.

(c) Airborne tests are not authorized.

c. False Alarms.

1. Caution should be exercised to prevent the inadvertent activation of ELTs in the air or while they are being handled on the ground. Accidental or unauthorized activation will generate an emergency signal that cannot be distinguished from the real thing, leading to expensive and frustrating searches. A false ELT signal could also interfere with genuine emergency transmissions and hinder or prevent the timely location of crash sites. Frequent false alarms could also result in complacency and decrease the vigorous reaction that must be attached to all ELT signals.

2. Numerous cases of inadvertent activation have occurred as a result of aerobatics, hard landings, movement by ground crews and aircraft maintenance. These false alarms can be minimized by monitoring 121.5 MHz and/or 243.0 MHz as follows:

(a) In flight when a receiver is available.

(b) Before engine shut down at the end of each flight.

(c) When the ELT is handled during installation or maintenance.

(d) When maintenance is being performed near the ELT.

(e) When a ground crew moves the aircraft.

(f) If an ELT signal is heard, turn off the aircraft's ELT to determine if it is transmitting. If it has been activated, maintenance might be required before the unit is returned to the "ARMED" position. You should contact the nearest Air Traffic facility and notify it of the inadvertent activation.

d. Inflight Monitoring and Reporting.

1. Pilots are encouraged to monitor 121.5 MHz and/or 243.0 MHz while inflight to assist in identifying possible emergency ELT transmissions. On receiving a signal, report the following information to the nearest air traffic facility:

(a) Your position at the time the signal was first heard.

(b) Your position at the time the signal was last heard.

(c) Your position at maximum signal strength.

(d) Your flight altitudes and frequency on which the emergency signal was heard: 121.5 MHz or 243.0 MHz. If possible, positions should be given relative to a navigation aid. If the aircraft has homing equipment, provide the bearing to the emergency signal with each reported position.

Bibliography

This list of publications, journals, and resources aided the author in researching for *Finding Carla*, in addition to the many interviews he conducted for this book. For those interested in further newspaper article information, the *Oregonian* is currently archived on various online search services such as GenealogyBank.com (fee-based, division of NewsBank, Inc.). Dozens of other newspapers around the country also carried similar accounts based on the Associated Press wire reports of the time.

Books

Collins, James C., *Good to Great* (HarperBusiness, 2001); "the Stockdale paradox" referenced on p.93

Department of Transportation/Federal Aviation Administration, *Federal Aviation Regulations* and *Aeronautical Information Manual* (ASA, 2015)

Gann, Ernest K., *Fate is the Hunter* (Simon & Schuster, 1961)

Jefford, Jack, *Winging It!* (Alaska Northwest Books, 2009); referenced on p.84

Mokler, R. J., *Aircraft Down: a personal account of search, survival, and rescue in the Canadian North* (Exposition Press, 1968); see p.95

Moore, Johnny, *I Must Fly!* (Sugarpine Aviators, 1998); referenced on p.44

Newspaper articles

Oregonian — Portland, Oregon
   3/18/1954, "Portland Airman To Face Charges"
   10/13/1956, "Flier Sleeps During Hunt"
   3/28/1959, "Landlord Arrested After Rent Dispute"
   4/5/1959, "Landlord Sentenced"
   12/15/1966, "Crew Seeks Missing Plane" (about pilot Art Ward)
   12/16/1966, "Search For Lost Pilot" (about pilot Art Ward)
   3/11/1967, "Klamath Crash Fatal to Four" (Flight 720 Crash)

3/13/1967, "Poor Weather In South – Portland Trio Feared Down in Light Plane"

3/14/1967, "Fresh Snow Slows Hunt for Airliner Crash Site"

3/14/1967, "Bad Weather Hampers Hunt For 3 Portlanders"

3/22/1967, "Teams Seek Lost Plane"

7/26/1967, "Search Party Finds Plane" (about pilot Art Ward)

10/3/1967, "4000 Hours Logged In Air By Lost Pilot"

10/7/1967, "CRASHFINDER Signal Urged"

10/13/1967, "Crash Case Closed But Many Queries Remain"; Leverett Richards

12/31/1967, "Annual Wrap-Up: Plane Crash Among Major Stories"

3/26/1968, "Radio In Every Plane Considered As Direction Aid To Searchers"

9/25/1969, "Dental Charts ID Remains of Al F. Oien"

9/26/1969, "Portland Plane Crash Victim Placed Glove In Tree to Mark Death Site"

Oregonian – Northwest Magazine

10/8/1967, "60 Days From March"; Art Chenowith

10/29/1967, "Why Didn't They Fight Harder To Live?"; Art Chenowith

Oregon Pilot's Association Newsletter — Beaverton Chapter (April 1967, "News and Views")

Port Angeles Evening News — Port Angeles, Washington

5/31/1971, "Emergency Radio Beacon was a Pretty Good Deal"

Redland Daily Facts — Redlands, California

3/14/1967, "Showers Drench West Coast, Snow Higher"

3/23/1967, "Snow Rain Sunshine in Weather Variety"

San Mateo Times — San Mateo, California

3/14/1967, "More Heavy Rains Due Tomorrow"

The Saturday Evening Post — Indianapolis, Indiana

1/7/1968, "Please Hurry, Someone"; Harold H. Martin

Websites

W.S. Hein & Co., Inc., *The Congressional Record:* HeinOnline.org

Rich Johnson, "Surviving a small plane crash": wilderness-urban-survival.blogspot.com

U.S. Department of Commerce, "U.S. Daily Weather Map" (March 9/10/11th, 1967): http://www.lib.noaa.gov/collections/imgdocmaps/daily_weather_maps.html

About the Author

Finding Carla is Ross Nixon's first book. His writing has appeared in *Alaska Magazine*. Born in Canada, he grew up in the United States on Washington's Olympic Peninsula, where he learned to fly. He has worked as a commercial diver, a police officer and served in the U.S. Navy Seabees as a welder. He makes his living as a commercial pilot, what he really wanted to be doing all along. An Alaskan resident since 1989, he enjoys flying, reading and writing. He lives in Anchorage with his wife Kate and their two cats, Madge and Edith.

Editor's Note

The surviving Oien family of Alvin Sr., Phyllis, and Carla, as well as others mentioned in the text have given the author permission to use statements in the family members' own words, some of which have appeared in previous newspaper and journal accounts of The Viking crash and search operations. The story the author tells in *Finding Carla* is based on his own recent interviews with the participants and he uses these with the permission of the interviewees. All the photographic and illustrative memorabilia in this book, including the Phyllis/Carla diary, belongs to the Oien family and is used with permission here.

Designer: Sarah Hager
Editing: Jackie Spanitz, Jennie Trerise